Edgar Fahs Smith, Harry Frederick Keller

Experiments arranged for students in general chemistry

Third Edition

Edgar Fahs Smith, Harry Frederick Keller

Experiments arranged for students in general chemistry
Third Edition

ISBN/EAN: 9783337215040

Printed in Europe, USA, Canada, Australia, Japan

Cover: Foto ©berggeist007 / pixelio.de

More available books at www.hansebooks.com

EXPERIMENTS

ARRANGED FOR

STUDENTS IN GENERAL CHEMISTRY

BY

EDGAR F. SMITH, AND HARRY F. KELLER,

PROFESSOR OF CHEMISTRY, UNIVERSITY OF PENNSYLVANIA.
PROFESSOR OF CHEMISTRY, CENTRAL HIGH SCHOOL OF PHILADELPHIA.

THIRD EDITION, ENLARGED, WITH 41 ILLUSTRATIONS.

PHILADELPHIA:
P. BLAKISTON, SON & CO.
1012 WALNUT STREET.
1895.

PRESS OF WM. F. FELL & CO.,
1220-24 SANSOM ST.,
PHILADELPHIA,

PREFACE TO THE THIRD EDITION.

This little work is designed as a guide for beginners in chemistry. The authors have found the course of instruction as arranged in the following pages productive of good; its object is not to dispense with the supervision of an instructor, but rather to assist him.

The present edition differs from its predecessors in the following points: many new experiments and new illustrations have been introduced, while not a few of the experiments described in former editions have been modified in accordance with the experience obtained in the daily use of the book with large classes. The text, too, has been subjected to careful revision.

Although reference is frequently made to Richter's "Inorganic Chemistry," fourth edition, any other text-book on the subject can be employed in its stead. The experiments have been collected from various sources, and no claim is made for originality.

CONTENTS.

NON-METALS.

CHAPTER		PAGE
I.	APPARATUS, MANIPULATIONS AND OPERATIONS,	9–10
II.	GENERAL PRINCIPLES,	11–12
III.	HYDROGEN,	13–17
IV.	CHLORINE, BROMINE, IODINE, FLUORINE,	17–25
V.	OXYGEN, SULPHUR,	25–35
VI.	NITROGEN, PHOSPHORUS, ARSENIC, ANTIMONY,	36–45
VII.	CARBON AND SILICON, BORON,	46–50

METALS.

VIII.	POTASSIUM, SODIUM [AMMONIUM],	51–58
IX.	CALCIUM, STRONTIUM, BARIUM,	58–60
X.	MAGNESIUM, ZINC,	60–63
XI.	MERCURY, COPPER, SILVER, GOLD,	63–69
XII.	ALUMINIUM, TIN, LEAD, BISMUTH,	69–73
XIII.	CHROMIUM, MANGANESE, IRON, NICKEL, COBALT,	74–83
XIV.	PLATINUM,	83–84
	APPENDIX,	85–86

NON-METALS.

CHAPTER I.

APPARATUS, MANIPULATIONS AND OPERATIONS.

(1) *The Bunsen Burner and the Blowpipe.*
1. Make a borax bead. 2. Dissolve a very minute quantity of manganese dioxide in it. 3. Heat in the oxidizing flame (?). 4. In the reducing flame (?). 5. Heat oxide of lead on charcoal in the reducing flame. 6. In the oxidizing flame.

(2) *Working with Glass Tubing and Rods.*
1. Cut various lengths of rods and tubing. 2. Round the sharp edges by softening and turning the ends in the lamp.

(3) *Construct a Wash bottle* (Fig. 1).

Fig. 1.

1. Soften a sound cork by rolling it under your foot on a clean floor. 2. Bore two parallel holes through it by means of a cork-borer. These perforations should be cylindrical and of less diameter than the glass tubes they are to receive. Use a rat-tail file in enlarging them. 3. Cut suitable lengths of glass tubing. 4. Draw the longer one to a fine point after softening in the flame. 5. Bend the tubes in an ordinary fish-tail burner, and round the sharp edges. 6. Fit the different pieces together.

(4) *Arrange some other form of apparatus for practice.*
(5) *The Balance.*

1. Weigh an object by placing it on the left-hand pan of the balance, and a weight judged about equal on the right-hand pan. Should the latter be found too heavy, replace it by the next smaller one; if too light, by the next heavier one. Then add systematically the smaller weights, until the needle points to the middle of the scale. The final adjustment is made with the rider. In adding or removing weights, the supports must always be raised.

(6) *Measuring Vessels.*

1. Measure off 10 cc. of water (*a*) in a cylinder, (*b*) in a burette, (*c*) in a pipette. Always read the *lower* meniscus. 2. Measure off similarly 50, 100 and 200 cc. of water, and determine their weight. 3. Measure the volume of 50 grams of oil of vitriol, and of 65 grams of muriatic acid. What are the specific gravities of these substances? Note the relation between weight and volume in the metric system.

(7) *Chemical Operations.*—Solution, evaporation, crystallization, precipitation, filtration, washing, and drying.

1. Place into a test-tube pure sodium carbonate, into another cobalt chloride, and add distilled water to each. Stir. What occurs? 2. To calcium carbonate, add water. Is there any change? Now add a little hydrochloric acid. What action has it? 3. Pour five cc. of strong hydrochloric acid upon powdered manganese dioxide; observe appearance and odor. Note, too, in each case, whether heat has any effect. Distinguish between chemical and mechanical solution. 4. Heat the cobalt chloride and the calcium carbonate solutions, each in a separate dish, on an iron plate, until the liquids are completely driven off (?). 5. Dissolve potassium chlorate in hot water, and allow to stand and cool (?). 6. To a portion of the cobalt chloride solution, add a solution of soda; boil. 7. Allow to settle and filter. 8. Wash the precipitate until pure water runs through the filter (?). 9. Heat the filter until perfectly dry.

CHAPTER II.

GENERAL PRINCIPLES.

(1) *Changes in Matter.*

1. Rub a glass rod with a piece of cloth, then touch particles of paper with it (?). 2. Through an insulated spiral of stout copper wire pass a current from two Bunsen cells. Place a piece of wrought iron—a nail will answer—inside the spiral, and bring iron filings in contact with it. What happens? Interrupt the current and note the result; repeat. 3. Heat a platinum wire in the non-luminous flame; is there any change? What is the effect of removing it?

Are the original properties of the substances in the above experiments altered, after the action of the forces of electricity, magnetism and heat has been stopped?

4. Mix intimately four parts, by weight, of finely powdered sulphur with seven parts of very finely divided iron (filings). Pass a magnet over a portion of the mixture. Another portion treat with carbon disulphide in a test-tube. Then heat the remaining portion in a tube over a gas-flame. Note carefully what occurs in each case. Powder the mass resulting from the last operation in a dry mortar. Can you extract from it any iron with a magnet, or dissolve out any sulphur with carbon disulphide? What inference do you draw from the facts observed? 5. Decompose water in Hofmann's apparatus by an electric current. The water should be acidulated with sulphuric acid to make it a conductor of electricity. A current from four to six Bunsen cells is required. To the gas, of which a larger volume has collected, apply a flame, and to the other a glowing spark at the end of a chip of wood (?) 6. Heat oxide of

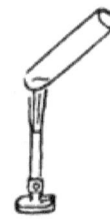

FIG. 2.

mercury in a tube of hard glass (Fig. 2.) (?)   7. Grind sulphur and mercury together in a mortar (?)   8. Heat sugar in a dry test-tube, at first gently, and then more strongly. Note color and odor.   9. Mix dry soda and tartaric acid in a mortar. Is there any action? What occurs when you add water? Point out in what respect the changes involved in experiments 1–3 differ essentially from those in 4–9. By what general names can you distinguish the two different kinds? With which does chemistry concern itself? Define *chemistry*.

Through what agencies have the results been obtained in experiments 4–9? Has any gain or loss of matter occurred in any of them?

(2) The products resulting from 5 and 6 cannot be further simplified, *i. e.*, decomposed into dissimilar substances. They are *elements*.* What are water and red oxide of mercury?

1. Dissolve in a little warm nitric acid, the black substance obtained by heating an intimate mixture of powdered sulphur and finely divided copper.

Evaporate the solution nearly to dryness, take up in water and filter. What remains on the filter? Place the filtrate in a beaker, dip the platinum electrodes of a battery (one or two Bunsen cells) into it (Fig. 3), and allow the current to act for 10 minutes. What do you observe upon the platinum foil, forming the negative pole? What changes have the copper and the sulphur undergone in this experiment?

(Study pp. 18–26, in Richter's Chemistry.)

(3) *Metals and Non-metals.*—(See Richter, p. 20.)

---

* The instructor should here develop the idea of *element* more fully.

CHAPTER III.

HYDROGEN.—H.

(1) Put several pieces of granulated zinc into a test-tube and pour dilute hydrochloric acid upon them. What occurs?

(2) Arrange the apparatus shown in Fig. 4. The flask should contain about 15 grams of zinc, and dilute sulphuric acid is poured through the funnel tube. When all the air in the apparatus has been displaced (*ask for precautions!*) collect six large test-tubes full of the gas over water.

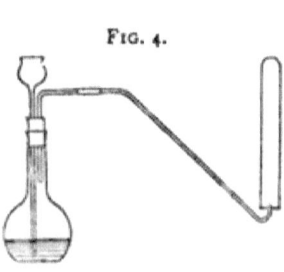

FIG. 4.

(3) What are its properties? Will it burn? Support combustion? Is it lighter than air?

Invert a test-tube containing one-third of its volume of water in a dish of water, and displace the water in the tube by hydrogen. What happens when the resulting mixture of air and hydrogen is brought in contact with a flame?

(4) Connect the stout copper wires of a Bunsen battery (three or four cells) with a loop of thin platinum wire. Introduce the incandescent wire into an inverted beaker containing hydrogen. What takes place?

(5) Over a small flame of burning hydrogen place a rather wide glass tube. Slowly lower the tube until a musical sound is heard. Explain this phenomenon.

(6) 1. To learn what becomes of hydrogen when it burns in air, arrange apparatus as in Fig. 5. The gas is led from the evolution flask A, into a bottle containing concentrated sulphuric acid, and then passes through a tube filled with pieces of calcium chloride. The gas which escapes is free from moisture. Burn it under a cold glass jar. What do you

obtain? 2. Fill a small flask with a mixture of one vol. of hydrogen and five vols. of air; cork; invert the flask several times to mix the gases; wrap a towel around it and bring its mouth to a flame. Result?

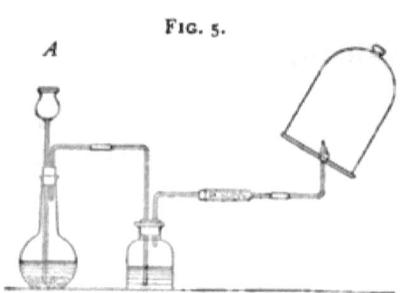

FIG. 5.

(7) *Hydrogen is not the only Product of the action of Sulphuric Acid upon Zinc.*

Pour some of the liquid remaining in the flask, in which hydrogen was generated, into a porcelain dish. Evaporate to about one-third of the original bulk; allow to stand several hours. You will now discover that the solution is full of colorless crystals. These are zinc sulphate or white vitriol—a salt, $ZnSO_4 + 7H_2O$. Write the equation of the reaction.

(8) *Determine the Weight of Hydrogen generated by a given Weight of Zinc.*

A piece of zinc (not more than .02 gram) is accurately weighed, and placed under a funnel in a beaker (Fig. 6). The latter is then nearly filled with water, so that the entire funnel is under the surface. A test-tube containing dilute sulphuric acid is lowered over the stem of the funnel. Hydrogen appears and collects in the tube. When all the zinc has disappeared,* transfer the tube containing the hydrogen to a larger vessel, holding water. Measure the volume of the gas by marking the tube where the inner and outer levels of water are even, and then weighing or measuring the

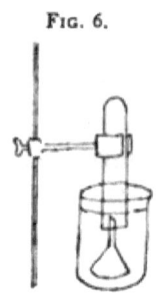

FIG. 6.

---

* This may be hastened by bringing a spiral of platinum wire in contact with the zinc.

HYDROGEN.

quantity of water that it will hold to that mark. Note the temperature of the water, and the height of the barometer.

The weight of the hydrogen is found by multiplying the volume by the weight of one cc., *i.e.*, .0000896 gram. Before this can be done, however, it is necessary to reduce the volume of the gas to 0° C. and 760 mm., as the above value has been determined under these conditions. If v = volume observed, t = temperature, and p = pressure, then

$$v_0 = \frac{v \times p}{(1 + at) \times 760}$$
and W = $v_0 \times$ .0000896. a = .003665.*

To calculate the quantity of zinc necessary to generate a unit of hydrogen, we say—

FIG. 7.

Wt of H : Wt of Zn :: 1 : x.

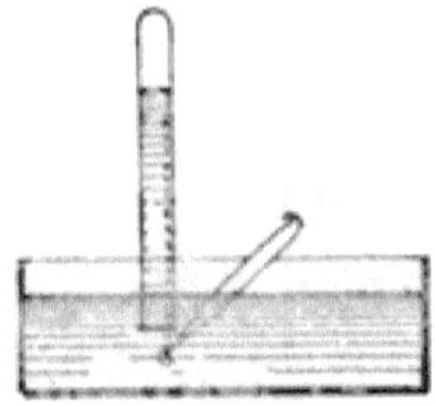

x here stands for the *equivalent weight* of zinc.

The equivalent weights of some other metals, such as iron, cadmium and magnesium, can be determined in the same manner. Magnesium gives the most satisfactory results.

(9) *Decompose water by electrolysis* and test the products.

(10) Wrap a small piece of sodium in paper and place it, with forceps, under the mouth of a test-tube filled with water, and inverted in water (Fig. 7) contained in a dish. Repeat this until the test-tube is filled with the gas. Test it for hydrogen.

What becomes of the metal ? Write the reaction.

(11) Construct the apparatus shown in Fig. 8.

Water is heated to boiling in the flask *A*, and the steam led over iron filings or wire, heated to redness in an iron tube.

---

* Tension of aqueous vapor is here neglected.

Collect the escaping gas over water. Test it for hydrogen Note its odor (?).

Is the iron changed? Equation?

(12) Into a tube of hard glass (six to eight inches in length)

FIG. 8.

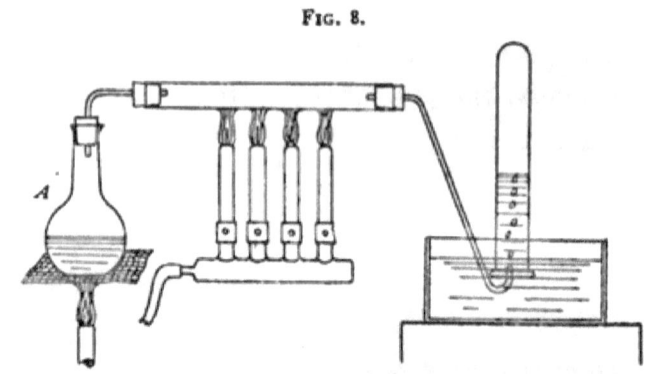

place a weighed quantity (one to two grams) of cupric oxide contained in a boat; connect the tube with a calcium-chloride tube of known weight (Fig. 9). Pass a current of hydrogen, dried by passing through concentrated sulphuric acid or cal-

FIG. 9.

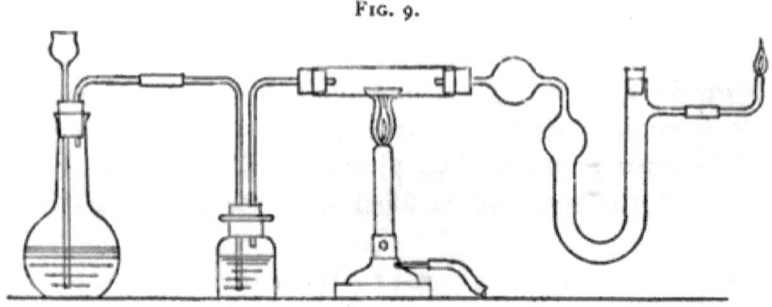

cium chloride, over the oxide of copper. When the air is completely expelled (?) apply a gentle heat to the part of the tube containing the oxide. Observe the glowing of the mass. When the change is complete, cool and determine

the loss in weight of the boat, and the gain in the calcium chloride tube. Explain the reaction.

Problems.—1. How much hydrogen can be obtained from zinc and 299 grams of sulphuric acid? 2. How much zinc and sulphuric acid are necessary to furnish 100 grams of hydrogen? 3. Suppose you have found that .015 gram of magnesium yields 15.2 cc. of hydrogen at 20° C. and 750 mm., what is the equivalent weight of that metal? 4. How many cc. of hydrogen can be obtained from two grams of sodium and water? 5. How many grams of water can be decomposed by 5 grams of iron; by how much is the weight of the latter increased? 6. 10 grams of cupric oxide will yield how much copper upon heating in hydrogen?

Give a brief summary of what you have learned about hydrogen?

(Study Richter, pp. 39–47.)

CHAPTER IV.

FIRST NATURAL GROUP OF ELEMENTS—CHLORINE, BROMINE, IODINE, FLUORINE.

CHLORINE.—Cl.

(1) Into a test-tube put manganese dioxide and concentrated hydrochloric acid. Note what happens both before and after heating.

(2) Use apparatus (shown in Fig. 10) for preparing larger quantities of chlorine. The manganese dioxide should be in the form of small lumps (not powder). Heat the mixture gently, pass the chlorine through a small quantity of water and collect it either by downward displacement or over *warm* water.

Write the reaction. How many atoms of chlorine are liberated? How many molecules?

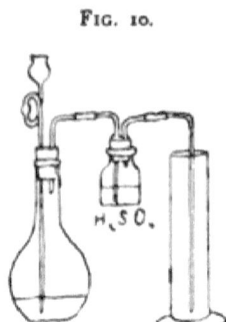

FIG. 10.

(3) 1. What is the normal condition of this element? 2. Is it lighter than air? 3. Is it inflammable? 4. Does it support combustion?

To obtain answers to these questions, fill a number of test-tubes with dry chlorine, and proceed as under hydrogen.

(4) Collect five large test-tubes full of the dry gas, and cover them with watch glasses.

Into 1 throw a little pulverized antimony.

Into 2 carefully introduce a piece of phosphorus.

Into 3 insert tissue paper moistened with oil of turpentine.*

Into 4 introduce colored flowers.

Into 5 pour a little litmus solution.

What are the results?

(5) Fill a test-tube with chlorine, and a second one of the same size with hydrogen. Bring the tubes together, mouth to mouth, and mix the gases by repeated inverting. Apply a flame to the open mouth of each tube (?).

(6) Invert a bottle filled with chlorine over water saturated with the same gas. What follows in the course of a few hours' exposure to sunlight? Can you account for results in experiments (4), (5), and (6)? Why should the gas be collected over *warm* water?

(7) *Determine the Weight of a Litre of Chlorine.*—Arrange apparatus as shown in Fig. 11.

In the evolution-flask place a mixture of equal weights of

---

* It is well, when the turpentine is old, to gently warm it, and then saturate the tissue paper.

salt and manganese dioxide. Add sulphuric acid, previously diluted with its own volume of water (pour the acid into the water!). Heat gently. Chlorine is evolved, and dried by passing through concentrated sulphuric acid, after which it is led into the perfectly dry flask c.* When this is filled, which you ascertain by the color of the gas in the neck, slowly withdraw the tube and cork the flask at once. Weigh the flask. Read the barometer and thermometer. Determine, also, the weights of the flask filled with air and with water.

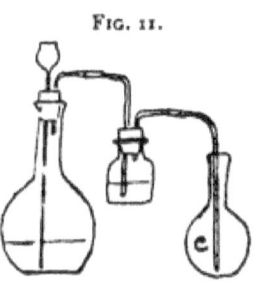

FIG. 11.

Calculation.—

Capacity of flask,	c
Temperature,	t
Pressure,	p
Flask filled with air,	w
" " chlorine,	w′
Wt. of a litre of air,	1.293 grm.
" " chlorine,	x

The weight of the air filling the flask is $\dfrac{c \times p \times .001293}{(1 + .00367\,t)\,760}$. The difference between this and w is the weight of the vacuous flask. Subtract this from w'. The remainder is the weight of the chlorine (W). Reduce the vol. of the chlorine to 0° C. and 760 mm. (see under hydrogen); it is $v_0 = \dfrac{a \times p}{(1 + .00367\,t)\,760}$, and the weight of 1 litre, $x = \dfrac{W \times 1000}{v^0}$.

How much heavier is one litre of chlorine than a litre of hydrogen?

---

* To prevent the diffusion of the gas into the air a plug of cotton should be placed in the neck of c.

Write the reaction involved in the above method for preparing chlorine.

Problems.—1. How many litres of chlorine can be obtained from one kilo of manganese dioxide and hydrochloric acid? 2. What weight of salt is required to prepare 100 litres of chlorine? 3. How many pounds of sodium sulphate and manganese sulphate will be formed in the preparation of 100 litres of chlorine gas? 4. Calculate the number of grams of chlorine that two litres of water will absorb, provided the latter takes up twice its volume of the gas? Write out your deductions from the above experiments on chlorine.

(Read Richter, pp. 49–52.)

HYDROGEN CHLORIDE.—HCl.

(1) Repeat the explosion of equal volumes of chlorine and hydrogen. Quickly cover the mouth of the flask, and immerse it under water. Does the latter rise? Put a drop of the liquid on the tongue and note the taste. Add some blue litmus solution. Is there any change?

(2) The product of the union of hydrogen and chlorine is a colorless gas. It is called *hydrogen chloride*. It is usually prepared by the action of sulphuric acid upon salt, thus:—

$$2NaCl + H_2SO_4 = 2HCl + Na_2SO_4.$$
or, better, $\quad NaCl + H_2SO_4 = HCl + NaHSO_4.$

The apparatus employed here is the same as that used in making chlorine (Fig. 10).

(3) *Determine the Properties of Hydrogen Chloride as under Hydrogen and Chlorine.*

What new property appears here? Fill a long dry glass tube with the gas, and quickly bring it into a basin containing water colored blue with litmus. What happens? What does hydrogen chloride gas yield on dissolving in water?

(4) In the preparation of hydrogen by the action of sodium

upon water, it was observed that the liquid became soapy to the touch, and acquired the property of turning red litmus blue. Prepare such a solution. To it add a few drops of litmus, and then a solution of hydrogen chloride (gradually) from a burette, until the blue color just begins to turn. Evaporate the resulting liquid to crystallization. Dissolve and recrystallize the product. It appears in cubes, and has the taste of common salt. It does not affect either red or blue litmus. We say it is *neutral* in reaction. The substance is chloride of sodium or common salt. What is a salt? An acid? A base? How can you obtain hydrogen chloride and chlorine from sodium chloride?

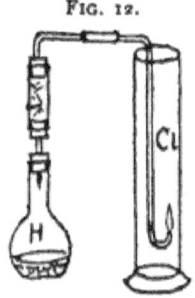

FIG. 12.

(5) *Burn Hydrogen in an Atmosphere of Chlorine, and Chlorine in Hydrogen.*

Generate chlorine as already described (p. 17) and collect it in a large cylinder. Into this introduce a burning jet of hydrogen (Fig. 12). Does it continue burning? What is the appearance of the flame? To show the combustion of chlorine in hydrogen arrange apparatus as in Fig. 13.

FIG. 13.

(6) To determine *the weight of a litre of hydrochloric acid gas*, proceed exactly as under chlorine.

(7) *Determine the Composition of Hydrochloric Acid Gas by Volume.*

1. Fill a perfectly dry and graduated tube with hydrogen chloride. Close the open end with the thumb, and opening the tube for a moment, quickly pour in about 10 cc. of sodium amalgam (see sodium, p. 53). Close the tube at once with the thumb, slightly moist, and shake well. Invert the tube in a large beaker of water, and remove the thumb. The amalgam will drop into the water, and the latter will

rush up into the tube, filling it nearly half full. Immerse the tube so that the water in it and that in the beaker are on the same level. This is done to measure the hydrogen under atmospheric pressure. Read the volume of the residual gas and measure also the volume of the mercury.

Calculation.—

Capacity of tube, ... a
Vol. of mercury, ... b
Vol. of hydrogen, ... c

$$c = \frac{a - b}{2}$$

(8) Add hydrochloric acid to solutions of silver nitrate; of mercurous nitrate; and of lead acetate. What do you observe in each case? Boil the precipitate formed in the lead solution with water. Cool, and note result.

BROMINE.—Br.

(1) Allow a drop of bromine to fall upon a heated watch glass; cover it quickly with a beaker. What is the color of the vapor? Dissolve one drop of bromine in each of the following solvents contained in test-tubes; water, alcohol, ether, carbon disulphide, and chloroform. Note the relative solubilities, and the color of each solution.

(2) 1. Pass chlorine through an aqueous solution of potassium bromide. What happens? 2. To a portion of the product add a few drops of carbon disulphide, and agitate the mixture; what is the result? 3. To another portion of the solution, containing free bromine, add a few drops of a starch emulsion.* Result?

---

* The starch emulsion for this purpose can be prepared as follows: One gram of starch is well ground in a mortar, with very little water, to creamy consistence. It is then poured into 200 cc. of boiling water. Allow to subside, decant the clear supernatant liquid and use it for the test.

(3) Devise a method for preparing bromine from potassium bromide.

(4) *Prepare Hydrobromic Acid.*—In a small flask cover five grams of red phosphorus with 10 cc. of water, and from a funnel, provided with a stop-cock, gradually allow 50 grams of bromine to run in.*

The gas is purified by conducting it through a U-tube, containing moistened pieces of phosphorus and glass (Fig. 14), and led into water to obtain the aqueous solution. How would you collect the gas?

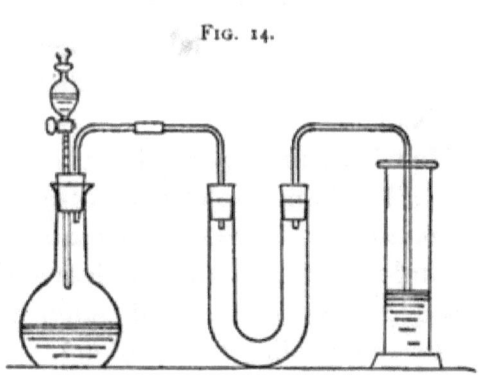

FIG. 14.

(5) Add hydrobromic acid to solutions of silver nitrate, mercurous nitrate and lead nitrate—do the resulting bromide precipitates differ much from the corresponding chlorides?

IODINE.—I.

(1) 1. Place an iodine crystal upon a warm plate, and note color of vapor. 2. Test the solubility of iodine in the same solvents as were used with bromine; what are the colors of the resulting solutions?

(2) 1. Pass chlorine into a solution of potassium iodide. Divide the resulting liquid into three parts. To one of these add about 5 cc. of ether and agitate (?). Shake a second portion with carbon disulphide or chloroform. To the remaining portion add some starch emulsion (?). 2. Repeat

---

* As it is rather difficult to weigh bromine upon a balance, calculate the volume corresponding to the weight given and measure out the same in a cylinder.

this experiment, substituting bromine water for the chlorine. Avoid excess of chlorine as well as bromine (?).

What conclusion do you draw from these experiments relative to the affinity of the halogens for potassium?

(3) Pass hydrogen sulphide gas (H_2S) into 50 cc. of water, and add powdered iodine till the brown color no longer disappears. Warm, filter (?) and distil the filtrate. The product is what?

How is gaseous hydriodic acid prepared?

(4) Precipitate solutions of silver nitrate ($AgNO_3$), mercurous nitrate ($HgNO_3$), lead nitrate ($Pb(NO_3)_2$), and mercuric chloride ($HgCl_2$), with potassium iodide. Note result in each case. Redissolve the lead iodide in water. What do you observe on cooling the solution?

FLUORINE.—Fl.

In what manner has this element been isolated?

(1) In a lead dish (or platinum crucible) place one gram of pulverized fluor spar ($CaFl_2$). Add concentrated sulphuric acid; cover the dish or crucible with a watch-glass coated with paraffin, through which characters have been drawn with a fine point. Heat gently for a few minutes.

What do you observe on removing the paraffin?

(2) Can you liberate fluorine from a fluoride?

Problems.—1. How much sodium bromide, sulphuric acid and manganese dioxide, are necessary to produce one cu. metre of bromine vapor at 20° C. and 745 mm.? 2. What per cent. of hydrogen iodide does a liquid contain, which represents a solution of 50 litres of the gas in one litre of water? 3. 10 grams of fluor spar will give what weight of hydrogen fluoride? 4. How much salt and sulphuric acid will be required to prepare six litres of muriatic acid of sp. gr. 1.17? What volume would the hydrogen chloride in these six litres occupy at 735 mm. pressure and 22° C.? 5. What is the per-

centage of hydrochloric acid in a solution of which 17 cc. dissolve exactly two grams of metallic magnesium? What is the volume of hydrogen liberated at 760 mm. and 0°?

CHAPTER V.

SECOND NATURAL GROUP OF ELEMENTS—OXYGEN, SULPHUR, SELENIUM, TELLURIUM.

OXYGEN.—O.

(1) *Preparation.*—1. Weigh the hard glass tube a (Fig. 15), and introduce a weighed quantity (about .5 gram) of red oxide of mercury. Ignite strongly; collect the liberated gas, and measure it. Weigh the tube with the residue. What are the products of the ignition?

(2) Prepare more of the gas, as follows: Mix equal parts of potassium chlorate and pulverized manganese dioxide; heat in a tube of hard glass or small retort. Collect the gas in bottles over water (Fig. 15).

Into No. 1 lower a piece of ignited sulphur on an iron spoon. Note result. Add water after the combustion (?).

Into No. 2 introduce a small piece of burning phosphorus (care!). Proceed as in No. 1.

Into No. 3 introduce ignited charcoal. Treat as before. Add now a few drops of blue litmus to the contents of each bottle. Any change?

Into bottle No. 4 introduce a fine watch spring, previously heated at one end and dipped into powdered sulphur. Result?

Is oxygen heavier or lighter than air? Has it color, taste, or odor? Will it burn? Does it support combustion?

What other methods can be used for preparing oxygen?

(3) *Determine the Weight of a Litre of Oxygen.*

Arrange apparatus shown in Fig. 16; a is a tube of hard glass, whose weight is known; it contains a weighed amount of potassium chlorate (about 0.3 gram). The bottle A is filled with water, b is a clip and d a beaker. The exit tube should be filled with water at the beginning of the experiment. Open the clip, heat a to bright redness, and receive the water displaced by the oxygen in d. When no more gas is evolved,

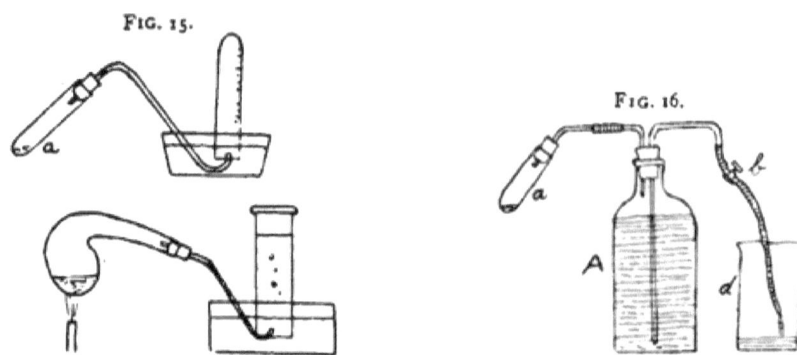

FIG. 15.

FIG. 16.

cool; allowing the rubber tube to dip under the water in the beaker. Some of the water will be drawn back into the bottle (?). Measure the volume of the water in d. Note the temperature of the air, and the height of the barometer. Weigh a, containing residue of potassium chloride (KCl).

Calculation.—

Weight of the tube,	a
Weight of potassium chlorate and tube,	b
Weight of potassium chlorate,	b — a
Volume of water collected,	v
Barometric pressure,	p
Temperature,	t
Aqueous tension at t,	p'
Weight of potassium chloride and tube,	c

$$V_0 = \frac{v \times (p - p')}{(1 + .00367\, t) \times 760} \quad \text{and} \quad x = \frac{(b - c) \times 1000}{V_0}$$

Dissolve the residual potassium chloride in water, and to its solution add nitrate of silver (?). How does potassium chlorate behave under like conditions?

(4) Give a summary of your work upon oxygen.

Problems.—1. How much oxygen, by weight and volume, can be obtained from 54 grams of mercuric oxide? 2. Heat will expel what volume of oxygen from 2.45 grams of potassium chlorate? 3. How much mercuric oxide is necessary to yield one cu. d. m. of oxygen? 4. How many times is oxygen heavier than hydrogen?

OZONE.—O_3.

(1) Pour water on clean pieces of phosphorus to half cover them; invert a large, clean jar over this and allow to stand for several hours. Test the air under the jar for ozone. For this purpose use paper impregnated with a mixture of starch paste and potassium iodide. What occurs?

(Read Richter, pp. 85–89.)

COMPOUNDS OF OXYGEN AND HYDROGEN.

WATER.—H_2O.

(1) Arrange the distillation apparatus (Fig. 17) and prepare about 100 cc. of distilled water. Note its taste and odor. Test it for chlorides with silver nitrate. Does it leave a residue upon evaporation? What action has it on litmus?

What is meant by the *hardness* of water? What is the understanding of *temporary* and *permanent* hardness?

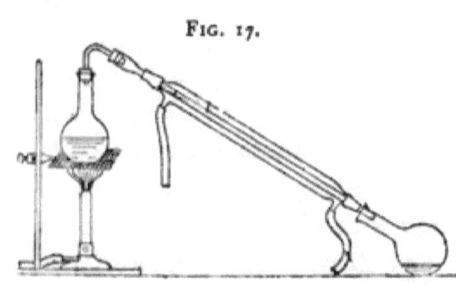

FIG. 17.

Determination of Hardness.—Dissolve one gram of pure calcium carbonate in 50 cc. of dilute hydrochloric acid. Evaporate this solution to dryness on a water-bath, and take up the residue in 50 cc. of distilled water. Each cc. of this solution will correspond to 0.001 gram of calcium carbonate. Next dissolve about 13 grams of castile soap in a mixture of 500 cc. alcohol and 500 cc. of water. Filter if necessary. Determine the strength of this solution so that one cc. of it will equal one cc. of the first solution or 0.001 gram $CaCO_3$. To this end remove 12 cc. of the first solution to a flask and dilute with water to 70 cc. Fill a burette with the soap solution and allow the latter to run into the lime water, one cc. at a time, shaking after every addition, until a lather is formed which lasts for about five minutes. Note the volume consumed. Dilute so that 12 cc. of the water require 13 cc. of the soap solution. One cc. soap is allowed for the distilled water. We can now consider the soap solution *standardized;* one cc. of it is equivalent to 0.001 gram of calcium carbonate ($CaCO_3$).

In determining the *hardness* in a natural water use 70 cc. and introduce the soap solution until a permanent lather is produced; deduct one cc. of the volume consumed, and the difference will represent the *hardness* of the water in terms of calcium carbonate. This result gives the total hardness. How would you ascertain the permanent and temporary hardness? What action has the soap upon the lime water?

Apply all these tests to a natural water (except rain).

(2) 1. Heat a little vegetable matter in a dry test-tube. 2. Heat fresh meat in the same manner. 3. Carefully heat crystals of zinc or copper sulphate in a test-tube. What happens in these experiments? 4. Expose clear crystals of sodium phosphate, on a watch crystal, to the air. 5. Do the same with pieces of calcium chloride. Results?

(3) *Determine the Quantitative Composition of Water.*

1. The composition of water by weight follows from the experiment of reducing oxide of copper described under hydrogen.

SECOND NATURAL GROUP OF ELEMENTS—WATER. 29

2. The relative volumes with which oxygen and hydrogen unite to form water, are determined either by analysis or synthesis. The former has been performed in electrolyzing water.

3. Fill a eudiometer (Fig. 18) with water. Through a rubber tube admit about 50 cc. of oxygen and then a like volume of hydrogen. (If the eudiometer is not graduated, mark these with rubber bands.) Close the open end with your thumb, leaving some air to serve as a cushion beneath it, and pass the spark. Remove the thumb, and pour in enough water to make the levels equal in both limbs. What is the

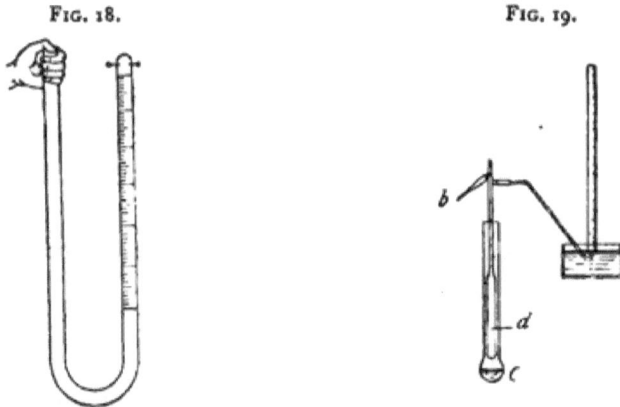

FIG. 18.   FIG. 19.

amount of the contraction? What is the residual gas? Test it.

(4) *Determine the Weight of a Litre of Steam.*—Use the apparatus of Victor Meyer shown in Fig. 19. C is a vessel containing aniline. A small glass tube is weighed and filled with water (not more than .02 gram). Heat the aniline to its boiling point, and continue heating until the temperature is constant (?). Now drop the tube containing the water through the side-tube b of the vessel d (the bottom of which should be protected with a layer of asbestos) and quickly re-cork. When the fall of water in the graduated tube ceases,

read the volume of gas, and note the temperature and pressure of the air.

The calculation is analogous to that used under oxygen?

(5) Perform experiment 2, p. 100 in Richter.

How many volumes of steam result from the combination of two volumes of hydrogen and one volume of oxygen?

How would you deduce the molecular formula of water from the preceding experiments?

HYDROGEN PEROXIDE.—H_2O_2.

(1) Add moist hydrated barium peroxide to cold dilute sulphuric acid. Filter. What does the filtrate contain?

(2) 1. Add a solution of potassium iodide, containing starch, to a portion of this liquid (?). Ferrous sulphate hastens the reaction. 2. Cautiously add a dilute solution of potassium permanganate to another portion (?). 3. To a third portion, add a few cc. of ether and a drop of potassium dichromate. Shake the mixture and observe the result.

Compounds of Oxygen and Chlorine.

(1) Make a dilute solution of caustic potash, and conduct chlorine into it until the latter is no longer absorbed. Treat one portion of the product with hydrochloric acid, and another with sulphuric acid. What results?

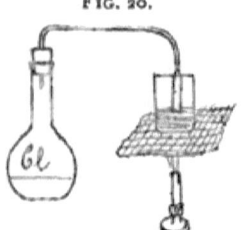
Fig. 20.

(2) Mix 10 grams of quicklime with 25 cc. of water. After the slaking is finished, conduct chlorine into the mixture until it is no longer absorbed. Add hydrochloric acid to one portion and sulphuric acid to a second portion. What is set free? Does it bleach?

(3) Pass chlorine into a hot *concentrated* solution of potassium hydroxide till it ceases to be absorbed (Fig. 20).

What separates upon cooling? Recrystallize the product from water. Will it give off oxygen upon heating? Try the action of hydrochloric acid upon a crystal. Allow a drop of concentrated sulphuric acid to fall upon a *small* crystal and warm gently (?). Care!

Observe carefully the behavior of potassium chlorate upon heating (?).

SULPHUR.—S.

(1) Place a few grams of powdered sulphur in a dry test-tube, and heat gradually. Observe and describe the changes which occur.

(2) Dissolve a little sulphur in carbon disulphide and allow to stand till the liquid has evaporated. What remains?

(3) Determine the specific gravity of sulphur. Water, previously boiled, is introduced into a flask provided with a mark (Fig. 21). It is essential that the neck of the flask should be narrow. Weigh the flask, then place an additional 10-gram weight upon the right hand pan of the balance and small pieces of sulphur upon the left-hand pan, until the pointer is again in the middle. Now introduce the sulphur into the flask. Carefully remove water above the mark and re-weigh the flask with its contents. The loss in weight will represent the weight of a volume of water equal to that of 10 grams of sulphur. The latter divided by the former is the specific gravity of the sulphur.

FIG. 21.

(4) Prepare the *monoclinic* modification of sulphur by melting about 10 grams of the ordinary variety in a covered Hessian crucible. Cool; and as soon as a solid crust has formed upon the surface, pierce it and allow the still liquid portion of the contents to run out. Note the shape of the crystals upon the sides of the crucible.

(5) To obtain the *plastic* variety, heat 20 grams of sulphur

in a small, round-bottomed flask until it boils, and pour it in a thin stream into cold water.

Test the solubility of the product in carbon disulphide. Preserve a portion of it for several days. Does it change?

(6) To a strong solution of yellow potassium sulphide, add hydrochloric acid. What are the properties of the separated sulphur?

Give a brief outline of the element sulphur; compare it with the previously studied elements.

SULPHUR AND HYDROGEN.

(7) *Hydrogen sulphide*—(H_2S)—is formed with difficulty from its elements, but is readily obtained by the action of acids upon sulphides, thus:—

$$FeS + H_2SO_4 = FeSO_4 + H_2S \text{ or } Sb_2S_3 + 6HCl = 2SbCl_3 + 3H_2S.$$

The apparatus to be used is the same as that employed in preparing hydrogen. The acid used should be *dilute*.

(8) What are the properties of hydrogen sulphide? Is it soluble in water? Does it burn? What are the products of its combustion? Hold a porcelain plate in the flame; what results?

(9) Expose a portion of the aqueous solution of hydrogen sulphide to the air. What causes it to become turbid and to lose its odor? Pass a current of the gas into strongly acid solutions of potassium chromate, potassium permanganate, and ferric chloride. Describe and explain the occurring changes.

(10) What action has hydrogen sulphide water on litmus?

(11) Pass hydrogen sulphide through solutions of the following salts, viz.:—copper sulphate, antimony chloride, lead nitrate, arsenious trioxide, and zinc acetate. Note results carefully.

Can sulphides be prepared in another manner? (See Chap. II, § 1.)

SECOND NATURAL GROUP OF ELEMENTS—SULPHUR. 33

(12) *Determine the Composition of Hydrogen Sulphide.*

Into a bent tube of hard glass, filled with mercury (Fig. 22), introduce dry hydrogen sulphide.* Place a piece of tin in the bent portion and heat it. Is the volume of the gas changed after the experiment, and what becomes of the piece of tin? Test the gas remaining in the tube. Do your results enable you to deduce the molecular formula of hydrogen sulphide. (See Richter, 4th ed., p. 110.) Trace the similarity between hydrogen sulphide and water. Write a summary of your experiments on hydrogen sulphide.

FIG. 22.

SULPHUR AND CHLORINE.

(13) *Sulphur Monochloride.*—1. Prepare this compound by conducting dry chlorine over molten sulphur. The product which distils over is collected in a dry test-tube, kept cold by immersion in ice water. 2. Redistil the product. Determine its boiling point in an apparatus similar to that pictured in Fig. 23. Note the color and odor of the product. Expose some of it to the air on a watch-glass. Add water to another portion contained in a test-tube. Note carefully what happens. Write the reaction, and examine for all the products.

FIG. 23.

SULPHUR AND OXYGEN.

(14) Burn sulphur in the air. Result? Burn pyrite (FeS_2) in the air. What are the properties of the resulting compound? It is *sulphur dioxide*—SO_2.

(15) Fit a small flask, as indicated in Fig. 24. Place

---

* The instructor should assist in performing this experiment.

5

copper turnings in it, then add sulphuric acid (strong) through the funnel tube. Warm. Is the product the same as that obtained in 14? Is it soluble in water? Has the aqueous solution the same properties as the gas? 2. Pass some of the gas into solutions of potassium dichromate and potassium permanganate acidulated with sulphuric acid. Repeat these experiments with the aqueous solution instead of the gas. What happens in each case? 3. Test the aqueous solution of sulphur dioxide with litmus. What is this solution commonly called? 4. Fill a dry jar with sulphur dioxide gas; introduce colored flowers. Note the result.

FIG. 24.

(16) What is the formula of *sulphurous acid?* How many series of salts can it form? How would you designate the different sodium salts? Add hydrochloric acid to a solution of sodium sulphite. What follows? Evaporate the solution to dryness and examine the residue. What is it? Write the reaction.

SULPHUR TRIOXIDE—SO_3. (Read Richter, p. 191.)

(17) *Sulphuric Acid*—H_2SO_4.—To prepare sulphuric acid arrange apparatus as in Fig. 25. The large flask, A, represents the lead chamber of the commercial method. The cork in it is provided with several perforations through which glass tubes pass; these serve to introduce the various gases. In flask a place copper turnings and concentrated sulphuric acid. When this mixture is heated what gas is evolved? Flask b contains dilute nitric acid and copper turnings. What is evolved when heat is applied? Boil water in flask c. Let air enter through d. e serves for the escape of the excess of gases (?). First introduce into A the products (?) from a and b, together with air—observe the frost-like deposition upon the vessel—what is it? Blow steam into A; what becomes of the crystalline sublimate? When 10 cc. to

20 cc. liquid have collected in A, interrupt the experiment, and study the product carefully. 1. Dilute a portion of it with water; what happens? 2. Test a portion of this diluted solution with litmus (?). 3. Another portion neutralize with sodium hydroxide and evaporate. What is the residue? Does it contain any sulphur? Prove this. 4. Add barium chloride to a third portion of the solution. What is the

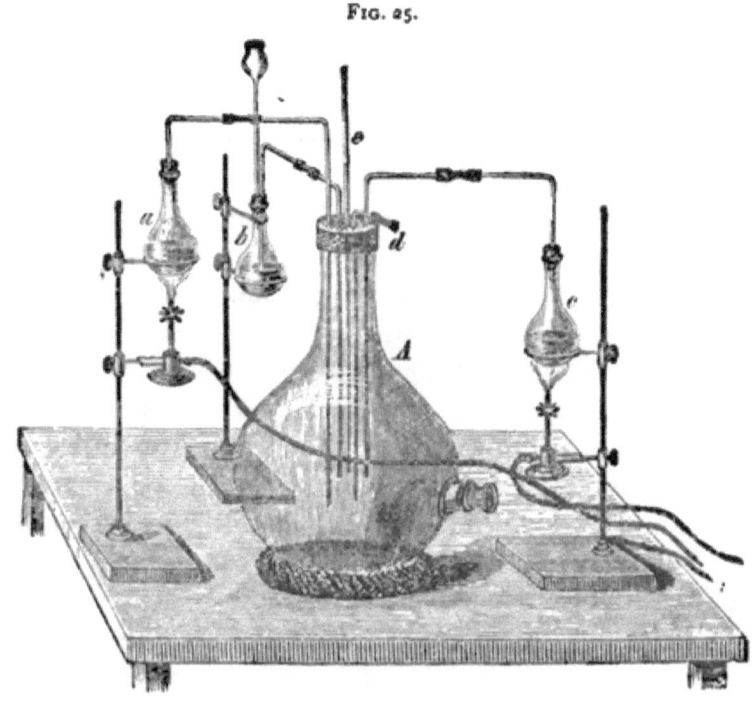

Fig. 25.

precipitate? Is it soluble in water or in hydrochloric acid? 5. What is the action of strong sulphuric acid upon wood or paper? Explain the cause of this action.

(18) How many series of salts can sulphuric acid form. Prepare ammonium sulphate, sodium sulphate, sodium hydrogen sulphate and copper sulphate. (Read Richter, pp. 189–200.)

CHAPTER VI.

NITROGEN GROUP—NITROGEN, PHOSPHORUS, ARSENIC, ANTIMONY AND BISMUTH.

NITROGEN.—N.

(1) *Preparation.*—1. In a dish swimming on water place a piece of phosphorus and ignite it; invert a beaker glass over it (Fig. 26). What becomes of the phosphorus? When the latter has ceased burning, restore the level of the water, and note the decrease in the volume of the air. Test the residual gas with a burning taper. 2. Heat gently in a small flask or retort a mixture of one part potassium nitrite, one part ammonium chloride, one part potassium bichromate, and three parts of water; collect the gas over water. Fill five bottles with this gas.

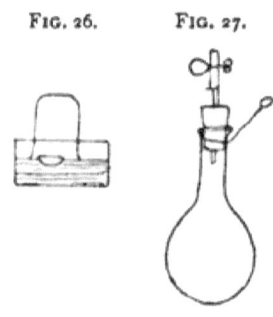

FIG. 26.   FIG. 27.

(2) Has it color, taste, odor? Does it burn or support combustion? Is the gas heavier than air? Does it unite readily with other elements?

(3) *Determine the Weight of a Litre of Nitrogen.*—A round-bottomed flask is fitted, as shown in Fig. 27. Pour about 30 cc. of water into it, and insert the rubber cork to the mark. Boil the water, while the clip is open, until all the air has been expelled from the flask. Steam should be allowed to escape for about five minutes. Now close the tube with the clip, and remove the flame. Cool and weigh the flask. Read the temperature and barometric pressure in the balance-room.

Connect the flask with the tube, b, of the aspirator, containing nitrogen, and arranged as in Fig. 28. The rubber

tube, *a*, is made to dip under water, and the clip is gradually opened, allowing nitrogen to enter the flask. Now raise the vessel containing the water into which the rubber tube dips, so that the water in it is at a higher level than that in the aspirator. Close the clip. Disconnect the flask and open the clip for a moment, to establish atmospheric pressure in the flask. Weigh. The calculation is identical with that given for oxygen.

What is the ratio between the weights of equal volumes of nitrogen and hydrogen?

(4) Is *air* a chemical compound?

How would you determine the weight of a litre of air?

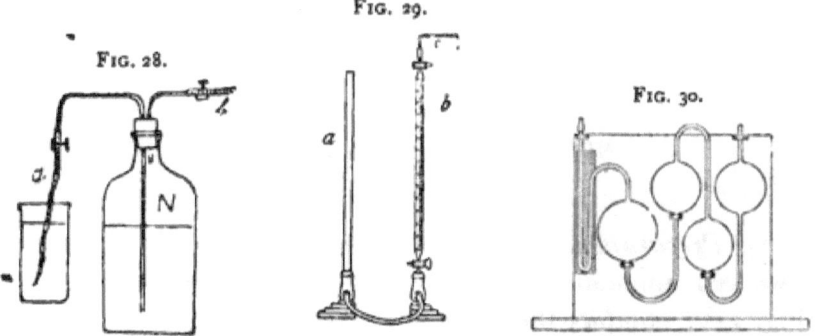

FIG. 28. FIG. 29. FIG. 30.

(5) 1. *Determination of the Oxygen in Air by the Pyrogallate Method.*—

At the atmospheric temperature and pressure measure off 50 cc. of air in the Hempel burette, shown in Fig. 29. Connect this at *c* with the capillary of a Hempel's compound pipette (Fig. 30) containing an alkaline solution of pyrogallate of potash. Open the stop-cocks and transfer the air to the pipette by raising the tube, *a*. When this is accomplished, and the capillary of the pipette is filled with water from *b*, close the stop-cocks again. Disconnect the apparatus. Shake the pipette for several minutes so as to bring gas and absorbent in intimate contact. Reconnect pipette and burette, and force the residual gas into the latter. Restore atmospheric pressure and read the volume. What does the loss represent?

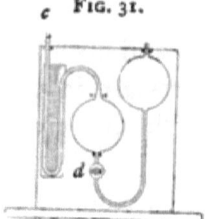
Fig. 31.

2. *Explosion Method.*—To 40 cc. of air contained in the burette add 40 cc. of pure hydrogen. Pass this mixture into the Hempel explosion pipette shown in Fig. 31. Close the stop-cock, *d*, and the clip, *c*, then connect the platinum electrodes with an inductor and pass a spark. What takes place? Measure the volume of the gas remaining. How much of the contraction was due to oxygen? What is the composition of the gas after the explosion?

(Study Richter, pp. 116–125.)

NITROGEN AND HYDROGEN.

AMMONIA.

(6) *Preparation.*—Heat an intimate mixture of finely powdered ammonium chloride and caustic lime in a flask (Fig. 32); conduct the evolved gas through a tube filled with small pieces of lime, and collect it in jars or test-tubes over mercury.

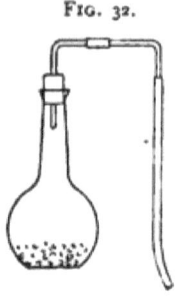
Fig. 32.

What is the object of the lime in the tube? Why can you not dry the gas by passing it through sulphuric acid or calcium chloride? Why should it be collected over mercury?

(7) Is ammonia gas combustible? Does it support combustion? 1. Arrange apparatus as shown in Fig. 33. *a* is a piece of glass tubing four to five cm. in diameter; its lower end is provided with a doubly perforated cork, carrying two tubes at right angles.

A slow current of ammonia is made to pass through the larger tube, while oxygen is introduced by means of the smaller tube. A plug of cotton serves to distribute the latter gas. Carefully regulate the flow of the gases and apply a lighted taper to the escaping ammonia. Note the peculiar appearance of the flame. 2. Heat concentrated ammonia water in a beaker until there is an abundant disengagement of gas, then conduct a rapid current of oxygen through the liquid, and lower a glowing spiral of platinum into the beaker (as in Fig. 34). What happens?

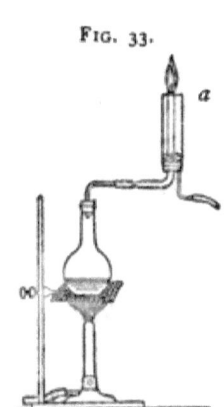

FIG. 33.

Note the odor of ammonia (caution?). Is it lighter than air? Soluble in water?

(8) *Prepare an Aqueous Solution of Ammonia.*
What are its properties?

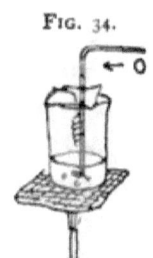

FIG. 34.

Add red litmus to some of the solution (?), and then neutralize carefully with dilute hydrochloric acid. Evaporate to dryness. Compare the product with the ordinary ammonium chloride. Test it for chlorine (?). Heat a little of it with sodium hydroxide (?). Heat another portion on a platinum foil (?).

(9) *Determine the Weight of a Litre of Ammonia.*

Fill a dry flask with the gas by *upward* displacement, and proceed exactly as under chlorine. What is the density of ammonia?

To determine the quantitative composition of ammonia, perform experiments 1 and 2 on pp. 130 and 131, in Richter.

Write out summary. (Read Richter, pp. 125-131.)

NITROGEN AND THE HALOGENS.

(10) Pour a saturated alcoholic solution of iodine into strong ammonia water. Collect the precipitate on a filter and wash it with water. Open the moist filter; tear it into small pieces and spread these on a board. After they have become dry, touch them with the end of a rod (?). *Ask for instructions!* (Read Richter, pp. 133-134.)

NITROGEN AND OXYGEN.

(11) *Nitrous Oxide*—N_2O.—1. Place about five grams of ammonium nitrate in a small retort, and heat gently. Collect the product over warm water. 2. Test it with a glimmering chip; 3. with burning phosphorus; 4. with burning sulphur. 5. Mix equal volumes of this gas and of hydrogen, and apply a flame. What other gas does it resemble in its properties? (Read Richter, pp. 215-216.)

(12) *Nitric Oxide*—NO.—1. Pour dilute nitric acid (sp. gr. 1.2) upon copper turnings contained in an evolution flask. Cool, and allow the red fumes, which form at first, to escape; then collect the colorless product over water. 2. What occurs when this gas comes in contact with the air? Is it the oxygen or the nitrogen of the air that acts upon the gas? 3. Apply the tests given under (11) to this gas (?). How can nitric oxide be distinguished from oxygen? 4. Fill a cylinder with nitric oxide, and add a few drops of carbon dioxide, shake well and bring a flame to the mouth of the vessel (?). 5. Pass a current of nitric oxide into a strong solution of ferrous sulphate. What occurs? After the solution has become saturated with the gas heat it to boiling (?). 6. Pass the gas into a solution of potassium permanganate (?).

(13) *Nitrogen Trioxide*—N_2O_3—(Read Richter, pp. 208-209.)

Nitrous Acid—HNO_2 (Richter, p. 209.)

(14) *Nitrogen Tetroxide*, N_2O_4, and *Dioxide*, NO_2.—1. Heat 10 grams of dry lead nitrate in a test-tube; condense the escaping vapors in a receiver, surrounded by a freezing mixture. What are the vapors, and what is the condensed liquid? Note the color. 2. What is the action of cold water, and of aqueous solutions of the alkalies upon nitrogen tetroxide? What do these reactions indicate in respect to the composition of this compound? (Richter, pp. 210-211.) 3. What is its action upon potassium iodide?

(15) *Nitrogen Pentoxide*, N_2O_5.—(Richter, p. 208.)

NITRIC ACID.—HNO_3.

1. *Preparation.*—In a retort heat a mixture of sodium nitrate and sulphuric acid in proportions corresponding to the equation (?):

$$NaNO_3 + H_2SO_4 = NaHSO_4 + HNO_3.$$

Collect the product in a cold receiver.

2. What are the physical properties of nitric acid? Color? Odor? Action on litmus (dilute with water)? 3. What action has it on indigo? Upon the skin? 4. Notice the effect of the acid upon the following metals: copper, zinc, iron, lead, tin. Write the reaction for each one. 5. Cover powdered sulphur with the acid, and warm (?). Dilute with water, filter, and test the liquid with barium chloride (?). 6. Add a few drops of nitric acid to a solution of ferrous sulphate (?); warm the solution (?).

Problems.—1. Required one cu. m. of nitrogen. How much air is to be deprived of oxygen; and how much phosphorus must be burned, if 62 parts of the latter unite with 80 parts of oxygen?

2. How much nitric acid, containing 46 per cent. of water, may be obtained from 1,700 grams of sodium nitrate, and how much water must be taken?

3. How many grams of ammonia will be absorbed by five litres of water, if the latter absorbs 500 times its volume of the gas? 4. Ten litres of water having absorbed 700 times their volume of ammonia, what are the least amounts of ammonium chloride and caustic lime necessary for producing this solution?

PHOSPHORUS.—P.

(1) 1. Determine the physical properties of the *active* and the *red* varieties. 2. Allow a small piece of the active variety to ignite in the air. Will the red variety do this? 3. Throw a small piece of the yellow variety into a jar of dry chlorine (?). Repeat with the red variety (?). 4. Bring a small dry piece of active phosphorus in contact with iodine (?). 5. Heat a flask containing a small piece of phosphorus and water until the former is melted, then pass a current of oxygen through a delivery tube into the melted phosphorus (?). *Care!* (Study Richter, pp. 134-137.)

PHOSPHORUS AND HYDROGEN.

(2) *Phosphine*—PH_3. To prepare phosphine arrange apparatus as shown in Fig. 35. A strong solution of caustic soda is placed in the flask, to which are added a few pieces of phosphorus. The air in the flask is now displaced by passing a current of coal gas through it. When this has been done close the clip, *a*, and gently heat the contents of the flask. What becomes of the gas as it escapes into the air? Write the reaction involved.

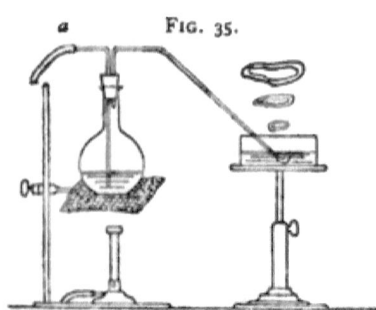
Fig. 35.

(Richter, pp. 137-140.)

Is there any similarity between phosphine and ammonia?

PHOSPHORUS AND THE HALOGENS.

(3) 1. Pass a current of dry carbon dioxide gas into a retort, the bottom of which is covered with dry sand. When all the air has been expelled, introduce some well-dried pieces of phosphorus, and replace the carbon dioxide by a stream of dry chlorine. Connect the neck of the retort with a Liebig's condenser, and collect the product in a receiver. It is phosphorus trichloride. What are its properties? Pour some of it into water (?).

2. Place a little phosphorus trichloride in a dry test-tube, and pass a stream of dry chlorine upon its surface. What is the result?

PHOSPHORUS AND OXYGEN.
(Richter, pp. 217-222.)

(4) 1. Prepare *phosphorus pentoxide*, P_2O_5, by burning a carefully dried piece of phosphorus under a dry bell-jar. 2. Drop a portion of the product into water (?).

(5) *Orthophosphoric Acid*, H_3PO_4; *Metaphosphoric Acid*, HPO_3; and *Pyrophosphoric Acid*, $H_4P_2O_7$.—How are these acids obtained? How many series of salts are derived from them? By what names would you distinguish the different salts?

1. Dissolve some disodium hydrogen phosphate, Na_2HPO_4, in water and test the solution with silver nitrate and ferric chloride. What do you observe in each case? 2. Dissolve *fused* sodium phosphate in water, and perform the same tests with its solution. 3. Heat salt of phosphorus ($NaNH_4HPO_4$) until it no longer effervesces; cool, crush the residue in a mortar, and dissolve it in water. How does this solution behave upon treating with the above reagents? 4. Acidify a portion of the last-named solution with acetic acid, and add a solution of albumen to it. Result?

(6) *Phosphorus Trioxide*—P_2O_3, and *Phosphorous Acid*—H_3PO_3.

Pour phosphorus trichloride into water. Evaporate the solution to syrupy consistency (?).

(Study Richter, p. 219.)

(7) *Hypophosphorous Acid*—H_3PO_2.

Heat pieces of phosphorus in a porcelain dish with a moderately strong baryta solution (see p. 42). When no more hydrogen phosphide is formed, cool, filter, and pass carbon dioxide into the solution until it shows a neutral reaction to litmus. Toward the end, the solution should be warmed. Filter and evaporate to suitable concentration. Hypophosphite of barium will crystallize.

How may the free acid be obtained from this salt?

ARSENIC.—As.

(1) Study the physical and chemical properties of this element. (Richter, pp. 143 and 144.) Are they analogous to those of phosphorus?

1. In a tube of hard glass heat a small piece of arsenic to redness. Result? 2. Heat arsenic with the oxidizing flame upon charcoal (?). 3. Dissolve powdered arsenic in strong nitric acid (?).

ARSENIC AND HYDROGEN.

(2) *Perform Marsh's Test for Arsenic.**

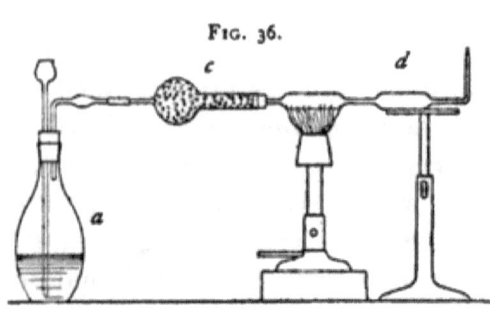

Fig. 36.

Arrange the apparatus shown in Fig. 36. To the mixture of zinc and dilute sulphuric acid contained in a, add a small portion of the solution to be tested for arsenic. The liberated gas contains hydrogen and arsine. It is passed through c, filled with

---

* Ask for instructions.

calcium chloride (?), and then through *d*, a tube of hard glass, contracted at several places. After all the air has been expelled from the apparatus, ignite the hydrogen. If arsenic is present it will burn with a bluish white flame, and white vapors will be given off. Hold a cold porcelain plate in the flame (?). Heat the tube *d*, as shown in the figure (?).

Great care must be exercised in performing this test, as the arsine gas is extremely poisonous!

ANTIMONY.—Sb.

(1) Study this element in the same manner as you studied arsenic. Distinguish between stibine and arsine.

1. Treat the metallic mirrors obtained in Marsh's apparatus with a freshly prepared solution of hypochlorite of sodium: Arsenic dissolves readily, while antimony is scarcely acted upon. 2. Heat a piece of the tube in which a mirror has formed, in the flame of the Bunsen burner. Dissolve the product in dilute, warm hydrochloric acid, and add hydrogen sulphide water (?). 3. Treat the spot formed upon a cold porcelain plate with yellow ammonium sulphide, and evaporate the solution at a gentle heat (?).

Problems.—(1) How much phosphorus can be obtained from 250 grams of bones? (See Richter, p. 135.) (2) 10 grams of phosphorus will give what volume of phosphine? (3) What is the weight of the product remaining, after evaporating a solution of 10 grams of arsenic in nitric acid.

CHAPTER VII.

CARBON GROUP—CARBON AND SILICON.

CARBON.—C.

(1) How many allotropic modifications of this element are known? What are their principal properties? In a rather wide tube collect dry ammonia gas (50–100 cc.) over mercury. With the aid of a forceps insert a piece of charcoal, which has just been ignited. What happens? 2. Boil a dilute litmus solution with powdered animal charcoal; filter. Result? 3. Substitute indigo for the litmus in the preceding experiment (?).

4. *Determination of the Composition of Coal.*

1. Volatile matter and coke. Weigh out two grams of powdered coal in a platinum crucible provided with a well-fitting cover. Heat with a large flame, until the escaping gases cease to burn between the lid and the crucible. A blast lamp flame is applied for a minute longer. Cool and weigh. Loss in weight represents the *volatile* matter. The residue is called *coke.*

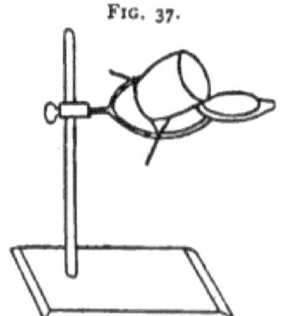

FIG. 37.

2. Ash. A second portion of coal (one gram) is gently heated over the Bunsen flame, until the volatile constituents are expelled. The heat is then raised and the lid of the crucible placed in the position indicated in Fig. 37. The residue is the *ash.*

(Read Richter, pp. 151–152).

CARBON AND HYDROGEN.

(2) *Methane* (Marsh gas)—CH_4.

1. *Preparation.*—Heat a dried mixture of sodium acetate

and sodium hydroxide in an iron tube.* Collect the gas over water. Note its color, odor, and taste. Does it burn? 2. Mix one volume of it with seven to eight times its volume of air and explode by applying a flame. (Ask for instructions!)

How would you determine the molecular weight of this compound?

(3) Make a eudiometric combustion of one volume of marsh gas with two volumes of oxygen, as described in Richter, p. 121.

(4) *Ethane*—C_2H_6. (Richter, p. 154.)

(5) *Acetylene*—C_2H_2. Light a Bunsen burner at the base and turn it down, so that the flame is small. Acetylene can be recognized, among the products of combustion, by its characteristic odor.

(6) CARBON AND THE HALOGENS. (Richter, p. 161.)

CARBON AND OXYGEN.

(7) *Carbon Dioxide*—CO_2.

1. *Preparation.*—Upon pieces of marble, contained in an evolution flask, pour dilute hydrochloric acid (1 HCl : 1–2 H_2O). Conduct the resulting gas through water and through concentrated sulphuric acid. It may be collected either by downward displacement of the air, or over mercury. 2. Note color, taste and odor of this gas. Is it soluble in water? How does its weight compare with that of air? Does it burn or support combustion? 3. Conduct a current of carbon dioxide into a solution of sodium hydroxide, evaporate the liquid, and test the residue for sodium carbonate (?). 4. To different portions of the sodium carbonate solution, add solutions of magnesium sulphate, barium chloride, lead nitrate and zinc sulphate.

---

* A hard glass tube will answer.

48   EXPERIMENTS IN GENERAL CHEMISTRY.

(Study Richter, pp. 230–235.)

(8) *Carbon Monoxide*—CO.

Preparation.—1. In a tube of hard glass heat zinc dust to faint redness, while conducting a slow current of carbon dioxide over it. In what respect does the product differ from carbon dioxide.   2. Heat crystals of oxalic acid with concentrated sulphuric acid in a flask, and wash the product with a sodium hydroxide solution. Write the reaction. Study the properties of this gas. (Richter, p. 235.)

(9) *Carbon Disulphide*—CS_2.

Perform some of the experiments indicated in Richter, p. 237.

(10) CARBON AND NITROGEN.

1. In a dry test-tube heat a nitrogenous carbon compound with a small piece of potassium. Cool and add water. Potassium cyanide is formed and can be tested with silver nitrate.

2. Convert a portion of the potassium cyanide into potassium sulphocyanide by evaporating with ammonium sulphide. Test with ferric chloride.   3. To a solution of ferrous sulphate add potassium ferrocyanide. What results?   4. What is the action of the ferrocyanide upon solutions of ferric salts?

(11) Study the nature of flame. Make the experiments described in Richter, pp. 156–161.

SILICON.—Si.

(1) *Preparation.*—Make an intimate mixture of one gram magnesium powder and four grams of finely powdered quartz-sand. Heat this to bright redness in a wide tube of hard glass. It is best to use the blast lamp for this purpose. The part of the tube containing the mixture should be rotated in the flame. The residue, after a few minutes' heating, is allowed to cool, and treated with water containing hydrochloric acid. The product consists of amorphous silicon and undecomposed quartz.   2. Test the action of the following

reagents upon silicon: sulphuric, nitric and hydrofluoric acids, potash solution and chlorine. (Read Richter, p. 162.)

SILICON AND OXYGEN.

(2) *Silicon Dioxide* (Silica, Quartz)—SiO_2.

1. Test its solubility in the various acids and alkalies. 2. Fuse a mixture of one gram of finely powdered quartz with four grams of sodium carbonate, in a platinum crucible. Dissolve the product in water. 3. To a portion of this solution add hydrochloric acid, and evaporate to complete dryness. Take up the residue with water and filter off the insoluble portion. 4. To another portion of the aqueous solution of the fusion add ammonium chloride (?). Make a bead of *salt of phosphorus;* bring a fragment of a silicate or of quartz into it, and heat in the blow-pipe flame for a few minutes (?).

BORON.—B.

(1) Preparation similar to that of silicon. What are its properties? Does it unite directly with other elements? Is it known in several allotropic modifications? What is the valency of this element?

(Read Richter, pp. 243 and 244.)

BORON AND OXYGEN.

(2) *Boric Acid*—H_3BO_3.

1. Dissolve borax in five parts of boiling water, add hydrochloric acid to acid reaction, and allow to cool. What crystallizes out of the solution? Dry some of the product by pressing it between filter paper. Test its solubility in water and in alcohol. What do you observe on igniting the alcoholic solution? Moisten a piece of turmeric paper with an aqueous solution of boric acid, and dry at a gentle heat. What happens?

Problems.—(1) How much carbon dioxide results from the combustion of 12 grams of carbon? (2) How much carbon dioxide will an indefinite quantity of calcium carbonate give, when acted upon by 4.666 grams of muriatic acid, containing 30 per cent. of pure hydrogen chloride gas? (3) How many cubic decimeters of carbon monoxide can be obtained from 90 grams of oxalic acid? (4) What amount of silica can be obtained from two grams of wollastonite ($CaSiO_3$)? (5) What is the theoretical quantity of boric acid obtainable from 15 grams of borax ($Na_2B_4O_7 + 10H_2O$)?

METALS.

CHAPTER VIII.

METALS OF THE ALKALIES—POTASSIUM, SODIUM, [AMMONIUM].

POTASSIUM.—K.

(1) *Preparation.*—Arrange apparatus as shown in Fig. 38. Into a tube of hard glass, introduce a porcelain boat containing about one gram of a mixture of 138 parts (one mol.) of *dry* (?) potassium carbonate and 72 parts (three at.) of magnesium powder. Pass a current of dry hydrogen over it, and after all the air has been displaced in the apparatus (?), light the escaping gas; heat the part of the tube surrounding the boat to incipient redness. Observe the brilliant metallic mirror which is formed, and drive it away from the boat by increasing the temperature: it is *potassium.* Note also the green color of the vapor and the violet coloration it imparts to the burning hydrogen. What is the residue left in the boat? Test its reaction with litmus (?).

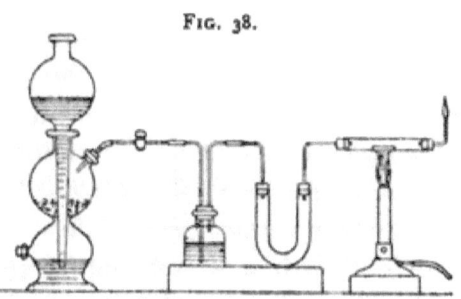

Fig. 38.

Formulate the reaction involved in this method of preparation.

(2) 1. Cut a piece of potassium with a knife, and observe

the color and lustre of the fresh surface. *Care!* 2. To ascertain whether the metal is fusible, heat a small piece of it in a stream of hydrogen. 3. Is it heavier or lighter than water?

(3) 1. Expose a thin slice of potassium to the air. What takes place? 2. Throw a *small* piece of it upon water (?). In this experiment it is advisable to use a tall beaker and to cover the same with a glass plate. 3. What is the action of the halogens upon potassium? *Ask for instructions.*

POTASSIUM AND OXYGEN.

(4) *Preparation of Potassium Hydroxide.*—In an iron vessel dissolve 50 grams of crystallized barium hydroxide, $Ba(OH)_2$, in 160 cc. of water. Cautiously add a hot concentrated solution of 20 grams of potassium sulphate until a sample of the supernatant liquid is no longer precipitated by either potassium sulphate or barium hydroxide. Filter rapidly through a plaited filter, and evaporate the solution in an iron or silver dish over a large flame. Continue heating the residue till it appears in a state of quiet fusion. *During this operation protect the eyes with a glass plate.* Now pour the product upon a clean iron surface, and while still warm put it into a bottle provided with a well-fitting stopper. Examine its fracture and color. Try its solubility in water and in alcohol. What is the reaction of the aqueous solution with litmus? What is an alkali?

Salts.

(5) *Potassium Chlorate.*—$KClO_3$. (See p. 30).

(6) *Potassium Nitrate.*—KNO_3.—To a hot concentrated solution of 20 grams of sodium nitrate add a solution of 18 grams of potassium chloride. Boil. What separates from the *warm* mixture? What crystallizes from the mother liquor on cooling? Recrystallize the latter product? Examine its crystalline form. Is it more soluble in hot than in cold water? Explain the method of preparation.

(7) Into a red-hot platinum crucible throw small portions of an intimate mixture of 10 grams of potassium nitrate and 1½ grams of charcoal powder. What takes place? Write the reaction. What is gunpowder?

Reactions.

(8) Use potassium nitrate for the following tests:—
1. Place a little of the salt upon the end of a clean platinum wire and introduce it into a non-luminous flame. What color do you observe? View the flame through a cobalt glass (?).
2. To the aqueous solution of the potassium salt add hydrochloric acid and boil. Concentrate by evaporation and add platinic chloride. What is the composition of the resulting precipitate? Try its solubility in hot and in cold water, also in alcohol. 3. To the concentrated solution of the salt add a saturated solution of tartaric acid; either at once, or on shaking, a white crystalline precipitate appears (?).

SODIUM.—Na.

(1) How is this metal usually prepared?
(2) Study its physical and chemical properties (Richter, p. 289). Wherein does it differ from potassium?
(3) *Prepare Sodium Amalgam.*
To 500 grams of dry mercury, contained in a Wedgwood mortar, add gradually 5–10 grams of sodium in thin slices. Perform this operation in a good draught chamber, as the union of the two metals is attended with the evolution of light and heat, and poisonous vapors are given off. Stir well with the pestle, allow to cool, and transfer the product to a well-stoppered bottle. What is its action on water or dilute sulphuric acid?

SODIUM AND OXYGEN.

(4) *Preparation of Sodium Hydroxide Solution.*
Add a little water to 10 grams of fresh quicklime contained

in an iron (or porcelain) vessel. Cover the latter, and in a second iron pot dissolve 25 grams of soda ash (Na_2CO_3), using about 100 cc. of water. Heat the solution to boiling; stir the quicklime—which should have broken up to a white powder—with enough water to form a thin paste (milk of lime), and add this gradually to the boiling liquid. Stir well with an iron wire; transfer the mixture to a bottle; cork, and allow it to stand. After the supernatant liquid has become perfectly clear, decant it by means of a glass siphon filled with water. It should be preserved in a tightly corked bottle (?). Test a few drops of the solution with barium chloride (?). What should the solution contain, and of what does the precipitate, from which it was separated, consist? Write the equation representing the reaction.

(5) *Determine the Amount of NaOH contained in the Solution.*

Measure off accurately 20 cc. into a porcelain dish; add a drop or two of phenolphthalein solution, and dilute with water. From a burette carefully add dilute hydrochloric acid until the red color has *just* disappeared. Read off the volume of the acid used; it is the exact quantity needed to neutralize the alkali:—

$$NaOH + HCl = NaCl + H_2O;$$

that is, 40 parts (one mol.) of sodium hydroxide require 36.5 parts (one mol.) of hydrochloric acid, and if we know the weight of the latter contained in the volume of the dilute acid consumed, a simple proportion will give the weight of the alkali in 20 cc. of the solution. The *strength* of the acid is determined as follows: In a porcelain dish, dissolve 1.06 grams of pure sodium carbonate, previously ignited and accurately weighed; add a little phenolphthalein, heat to boiling and introduce acid from the burette until the liquid *remains* colorless after continued boiling. The carbonate is then exactly neutralized:—

$$Na_2CO_3 + 2HCl = 2NaCl + CO_2 + H_2O.$$

It takes, therefore, 73 parts of hydrochloric acid for 106 parts of sodium carbonate. Suppose, now, 20 cc. of the acid had been used to decolorize the indicator, then one cc. would equal $1\frac{0.6}{20} = 0.053$ gram of sodium carbonate or 0.0365 gram of hydrochloric acid. The latter number is the *standard* or *strength* of the dilute acid.

The phenolphthalein takes no part in these reactions; it merely *indicates* by its change of color the complete neutralization of the alkali. Why is it necessary to boil the solution when the acid is standardized with a carbonate?

Salts.

(6) *Sodium Chloride*—NaCl.

Purify Common Salt.—Grind 50 grams of salt in a mortar with 150 cc. of water. Filter into a beaker, and conduct hydrochloric acid gas into the solution, as shown in Fig. 39. Pure salt separates out. Collect it on a platinum cone, remove the liquid with the aid of a filter pump, and dry the salt by warming it in a porcelain dish, while stirring it with a glass rod.

(7) *Sodium Carbonate.*—Na_2CO_3.

Recrystallize some of the commercial carbonate. Heat a portion of the product in a porcelain dish? What do you observe?

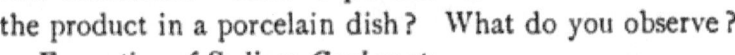
FIG. 39.

Formation of Sodium Carbonate.

(1) Prepare a finely divided mixture, consisting of six parts of dehydrated sodium sulphate (Glauber's salt), four parts of chalk, and one part of carbon. Heat this upon a platinum foil over a blast lamp until it fuses. After cooling, place the foil in a porcelain dish, add a little water, and heat to boiling. Filter the solution into a test-tube, and to a portion of the filtrate add hydrochloric acid. What happens? It indi-

cates the presence of what? In a second portion of the filtrate add hydrochloric acid. In the mouth of the test-tube suspend a strip of filter paper previously moistened with lead acetate. What happens? Explain. This experiment may be said to illustrate what technical process?

(2) Make a cold saturated solution of ammonium carbonate by shaking the finely divided commercial salt repeatedly with cold water. A saturated sodium chloride solution is prepared in the same manner. Conduct a current of carbon dioxide into the ammonium carbonate solution, and occasionally shake the vessel containing the solution. In the course of half an hour mix the two liquors, shake the mixture vigorously, run in the carbon dioxide again, and alternate these operations until a crystalline precipitate separates. What is it? Dissolve a portion of it in water, add hydrochloric acid. Dip a clean platinum wire into the solution and hold it in a colorless Bunsen flame. What name is given to the technical process based on this principle?

Reactions.

(8) Use the purified chloride for the tests. 1. What color do sodium salts give to the flame? 2. Mix a drop of the aqueous solution with 10 drops of a platinic chloride solution on a watch-glass. Evaporate *very carefully* to a small volume. On cooling, a red-colored salt crystallizes out in long monoclinic needles (?). Is it soluble in water? in alcohol? 3. Are there any salts of sodium which are not soluble in water? Can compounds of sodium be precipitated by any reagent?

AMMONIUM.

(1) What is the composition of ammonium? Can it be obtained in a free state? (See Richter, p. 299.) To a warm concentrated solution of ammonium chloride, contained in a large dish, add sodium amalgam (see p. 53). What

occurs? Hold a piece of reddened litmus paper over the dish (?).

(2) Dissolve commercial sal ammoniac in a little water, add ammonia in slight excess, warm, filter if a precipitate is formed, and evaporate to crystallization; stir constantly. Ammonium chloride is thus obtained in the form of a fine powder.

Reactions.

(3) 1. On a piece of platinum foil heat successively small portions of the chloride, the sulphate, and the nitrate. What occurs in each case? 2. Mix a little ammonium chloride with burnt lime in a small mortar. Note the odor of the escaping gas and its reaction with litmus. 3. Heat a small portion of ammonium chloride with a caustic soda solution. What is given off? Explain the action of strong bases upon ammonium salts. 4. Add platinic chloride to a solution of ammonium chloride. Result? 5. To a concentrated solution of the ammonium salt add tartaric acid and shake the mixture (?). 6. Do compounds of ammonium impart a color to the flame?

Compare the metals of the alkalies with each other. How can the compounds of potassium, sodium, and ammonium be distinguished?

Problems.—1. How much potassium nitrate is theoretically obtainable from two kilos of Chili saltpetre of 97 per cent., and what amount of sylvite containing 98 per cent. of potassium chloride is required? 2. Suppose that 75 cc. of dilute nitric acid were required to saturate 50 cc. of a potash lye; further, that 10 cc. of the acid neutralized 1.06 grams of sodium carbonate, what amount of caustic potash would the lye contain? 3. In the valuation of a pearl ash (impure K_2CO_3), 29.1 cc. of a sulphuric acid were used to neutralize five grams of the sample; the acid contained 98 grams of sulphuric acid per litre ; calculate the percentage of impurities in the product. 4. Required the minimum amount of marble that should be burnt to liberate the ammonia from 50 grams of ammonium nitrate.

CHAPTER IX.

METALS OF THE ALKALINE EARTHS—CALCIUM, STRONTIUM, BARIUM.

CALCIUM.—Ca.

CALCIUM AND OXYGEN.

(1) 1. Ignite two grams of powdered marble in a platinum crucible to the highest temperature obtainable with the aid of the blast lamp. Continue this for 15 minutes, occasionally stirring the mass with a platinum wire ; what is the residue ? Explain the reaction. 2. Add about five cc. of water to the product. What do you observe ? Test the reaction of the product with litmus paper.

(2) 1. *Prepare Lime-Water.*—To the slaked lime obtained from twenty grams of quicklime (see p. 54) add one litre of water ; transfer the mixture to a bottle. Cork tightly, shake and allow to stand. When the solution has become clear, draw it off by means of a siphon ? What does it contain ? Of what does the undissolved portion consist ? 2. Place a portion of the lime-water on a watch glass and expose to the air (?). 3. Through a second portion blow air from your lungs (?). 4. Conduct a stream of carbon dioxide through a third portion and observe carefully the successive changes. Explain them. 5. What takes place upon boiling the clear solution which is obtained as the final product in the preceding experiment ?

Salts.

(3) *Calcium Chloride.*—$CaCl_2$.

1. Evaporate some of the spent acid of a carbon dioxide generator to dryness. What is the residue? 2. Expose a little of the salt to the air (?). 3. What use have you made of calcium chloride previously? 4. *Prepare porous calcium chloride* ($CaCl_2 + 2H_2O$). Dissolve the residue obtained in 1 in lime-water, filter, and neutralize exactly with hydrochloric acid. Evaporate the filtrate to dryness in a porcelain dish, and heat the residue for some time on the sand-bath. The solution of the product must show a neutral reaction.

(4) *Calcium Hypochlorite.*—$Ca(ClO)_2$. (See p. 30.)

(5) *Calcium Sulphate.*—$CaSO_4$.

1. *Carefully* heat a few grams of gypsum in a porcelain dish until the water of crystallization is completely expelled. Pulverize the residue. What happens when it is made into a paste with water and allowed to stand?

Reactions.

Use the pure calcium chloride for the following tests:—

1. Introduce a small portion of the salt into the Bunsen flame by means of a platinum wire (?). 2. To the aqueous solution add ammonium carbonate. Result? 3. To another portion add dilute sulphuric acid. What is the composition of the precipitate? Why does it not form in very dilute solutions? 4. Add ammonia water and ammonium oxalate to the filtrate from the calcium sulphate.

STRONTIUM.—Sr.

Reactions.

1. What color is imparted to the Bunsen flame by compounds of this element? 2. Add a gypsum solution to the solution of a strontium salt (?).

BARIUM.—Ba.

Reactions.

1. Observe what color barium compounds give to the flame. Moisten the sample with hydrochloric acid before heating it (?). 2. To a portion of the aqueous solution of the chloride add ammonium carbonate. What results? 3. Add dilute sulphuric acid to a second portion (?).

Point out how the elements of this group may be distinguished (*a*) from those of the preceding group; (*b*) from each other.

Problems.—1. How much nitric acid of 20 per cent. will effect the solution of one gram of Iceland spar ($CaCO_3$)? How much carbon dioxide is given off, and what volume would it occupy at 20° C. under a pressure of 750 mm.? 2. Suppose .5 gram of sulphur were dissolved in nitric acid, what quantity of barium chloride must be added until it ceases to produce a precipitate? 3. One gram of a mineral consisting of the carbonates of calcium, strontium, and barium, in the proportion of their molecular weights, will leave what weight of the mixed sulphates on treating and evaporating with an excess of sulphuric acid?

CHAPTER X.

MAGNESIUM GROUP—MAGNESIUM, ZINC, CADMIUM.

MAGNESIUM.—Mg.

(1) Examine the metal in the forms of ingot, ribbon and powder. Note its color, lustre and specific gravity. 2. Introduce a piece of the ribbon into the flame with the forceps (?).

What is the product? 3. Treat a piece of the ribbon with dilute sulphuric acid. Reaction?

Salts.

(2) *Magnesium Chloride.*—$MgCl_2$.

Prepare the ANHYDROUS *salt.*—Dissolve about 50 grams of the crystallized (?) chloride and 50 grams of ammonium chloride in as little water as possible. Evaporate to dryness in a porcelain dish. Reduce the mass while hot to small pieces in a mortar, dry it carefully, so as to remove every trace of moisture. It is best to do this by heating small portions of the material in a porcelain crucible until it no longer sinters. A small sample should not give off moisture when heated in a dry test-tube. Be careful also to prevent re-absorption of moisture. Quickly transfer the warm powder to a platinum crucible provided with a well-fitting cover. Heat, at first gently, to expel the ammonium chloride, then increase the temperature until the mass is in a state of quiet fusion. It is the anhydrous salt which, being extremely hygroscopic, should be preserved in a tightly-stoppered bottle. It should dissolve in water to a *clear* liquid.

Why cannot the anhydrous chloride be obtained by evaporation of the aqueous solution?

(3) *Magnesium Sulphate.*—$MgSO_4 + 7H_2O$.

Recrystallize some of the commercial salt. What is the form of the crystals? Taste?

(4) *Reactions.*

1. Heat a portion of the sulphate or chloride on a platinum wire in the Bunsen flame; moisten with cobalt nitrate solution and heat again. A *pink-colored* mass results. 2. Add some caustic soda to a little of the solution of the chloride (?). The resulting precipitate dissolves on addition of an ammonium salt (?). 3. Mix a second portion of the chloride solu-

tion with ammonia water and ammonium chloride, add disodium hydrogen phosphate and agitate the liquid. What is the composition of the precipitate? Examine it with the aid of a lens.

ZINC.—Zn.

(1) How is this metal obtained from its ores?

(2) Study the physical and chemical properties of zinc (see Richter, p. 320). 1. Treat a small piece of *pure* metal with dilute sulphuric acid (?). 2. Repeat this experiment, substituting the impure commercial metal. What difference do you observe? What causes it?

(3) *Granulate Commercial Zinc.*—Melt 100 grams of the metal in a well-covered Hessian crucible. The blast lamp may be used for this purpose, but it is better to perform the operation in a wind furnace. The crucible is then removed from the source of heat, and allowed to cool until the molten metal no longer takes fire when the cover is lifted. Pour the metal, in a thin stream, into a pail filled with cold water. Drain the product and dry at a moderate heat.

Salts.

(4) *Zinc Sulphate.*—$Zn\,SO_4 + 7H_2O$. (See p. 14). 1. Prepare some of this salt and recrystallize it carefully from water. 2. Examine the crystals. What other salt have you prepared that exhibits similar forms? Is there any analogy in the composition of the two salts?

Reactions.

(5) 1. Heat a small piece of zinc on charcoal in the oxidizing flame (?). 2. Moisten the incrustation obtained with a drop of cobalt nitrate, and heat again. Result? 3. To a solution of zinc sulphate add ammonium sulphide. What is the color of the precipitate? Try its solubility in dilute hydrochloric acid and in $HC_2H_3O_2$ (acetic acid). 4. Study the

action of caustic alkalies, *e. g.* caustic soda, upon the zinc solution.

How could you distinguish between zinc and magnesium? What differences are there between this and the preceding groups?

Problems.—1. What is the strength of a sulphuric acid of which 20 cc. dissolve exactly .048 gram of magnesium? 2. Suppose it was found that one gram of zinc gave with sulphuric acid, 325 cc. of hydrogen at 16° C. and 755 mm., and, further, that .369 gram of magnesium produced the same amount of the gas. Knowing the atomic weight of magnesium to be 24, and remembering that the two sulphates are isomorphous, how is it possible to deduce the atomic weight of zinc from the data given?

CHAPTER XI.

MERCURY, COPPER, SILVER, GOLD.

MERCURY.—Hg.

(1) Study the physical and chemical properties of the metal. Wherein does it differ from the other metals?

MERCURY AND OXYGEN.

(2) *Mercuric Oxide.*—HgO.

How is this substance prepared? What is its behavior on heating?

Mix a little powdered sulphur with dry sodium carbonate and mercuric oxide. Ignite the mixture in a dry test-tube. Extract the residue with water, filter, acidify with hydrochloric acid and add barium chloride. What has become of the oxide of mercury in this experiment?

Salts.

(3) *Mercurous Nitrate.*—$HgNO_3$.

An *excess* of metallic mercury (use 10–15 grams) is treated in the cold with moderately strong nitric acid until the formation of crystals is no longer noticeable. Redissolve the crystals by warming, filter, and allow to crystallize.

To prepare a solution of the salt take it up with water acidulated with nitric acid (?).

(4) *Mercuric Chloride.*—$HgCl_2$.

Dissolve about five grams of metallic mercury in aqua regia. Evaporate to dryness on a water bath. Place the residue into a small dry flask, cover the latter with a watch-glass, and heat cautiously on a sand-bath. What is the sublimate formed in the upper part of the flask? Dissolve it in four parts of boiling water and allow to crystallize.

Reactions.

(5) *Mercurous Compounds.*—Use the solution of the nitrate. 1. Add a few drops of hydrochloric acid to two or three cc. of the solution. What takes place? Filter, and add ammonia water to the precipitate (?). 2. Add stannous chloride to another portion of the nitrate solution (?). 3. In a third portion immerse a strip of copper foil. Examine the stain on the metal; is it changed when you hold it in the flame?

(6) *Mercuric Compounds.*—The chloride will answer for the tests.

1. Pass hydrogen sulphide through a dilute solution and observe the gradation of colors through which the precipitate passes. What is the final product? 2. Add stannous chloride, drop by drop, to the mercury solution. Explain the changes which occur.

COPPER.—Cu.

1. *Preparation.*—Ignite the pure oxide in a current of dry hydrogen (see p. 16). Examine the color and the lustre of

the product; test its solubility in hydrochloric acid, sulphuric acid (both strong and dilute), and nitric acid. Write equations representing the reactions.

Salts.

(2) *Copper Sulphate.*—$CuSO_4 + 5H_2O$.

To 12 grams of copper in a flask add 45 grams of concentrated sulphuric acid, and heat. When the metal has completely disappeared and the gas (?) ceases to be given off, allow to cool, place the white crystalline residue (?) into a porcelain dish, rinse the flask with hot water. Now add a few drops of nitric acid to the hot water solution, and filter. From the filtrate the sulphate crystallizes on standing. Recrystallize the product.

Does this salt suffer decomposition on exposure to the atmosphere? Heat a small quantity in a porcelain crucible, first moderately, then more strongly (?).

(3) *Sulphate of Copper and Potassium.*—$CuK_2(SO_4)_2 + 6H_2O$.

Prepare solutions of 10 grams of blue vitriol and seven grams of potassium sulphate, both saturated at 70°. The latter should also contain a few drops of sulphuric acid. Mix the solutions; on cooling the double salt separates in whitish-blue crystals. Examine their form.

Reactions.

(4) Use either of the salts you have prepared.

1. Mix a little of the salt with sodium carbonate, and heat on charcoal in the reducing flame (?). 2. Make a borax bead and dissolve a minute quantity of a copper compound in it. What color does it give (*a*) in the oxidizing flame? (*b*) in the reducing flame? (*c*) when the bead is reduced with a small piece of tin? 3. Through a dilute copper solution pass hydrogen sulphide. Is the resulting precipitate soluble in

hydrochloric acid or in nitric acid? 4. Add ammonia, drop by drop, to the solution. What changes do you observe? 5. To a portion of the very dilute solution add potassium ferrocyanide (?).

(5) To a solution of copper sulphate in a porcelain dish add a small piece of zinc. Allow to stand over night. Note the result. Has the zinc disappeared? Does the solution contain any of this metal? In what form? Where is the copper?

(6) Repeat the experiment, weighing the copper sulphate (.1 gram) and the zinc (.5 gram). Add hydrochloric acid in quantity sufficient to insure the entire solution of the zinc, collect the copper on a filter, wash with alcohol, dry, heat gently and weigh it in a porcelain crucible. The filtrate should be colorless.

Compare the weight of the metallic copper obtained with that of the zinc employed (?). How does the found copper accord with the calculated amount of that metal in .5 gram of $CuSO_4.5H_2O$?

Repeat the experiment, using cadmium in place of zinc. Compare the weights of the metals as before. What deduction can you make?

Dissolve one gram of pure copper sulphate in 250 cc. of water; to this add one gram of ammonium sulphate and 10 cc. of ammonia water (specific gravity 0.96). Transfer this solution to a platinum dish, connect the latter with the cathode (?) of a battery consisting of from four to six gravity-cells, and in the solution suspend a flat-spiral of platinum in connection with the anode (?) of the battery. As the current continues to act what changes occur? How will you determine when the reaction is at an end? When this point has been reached, add sodium acetate to the liquid, interrupt the current, pour out the solution and wash the deposit of copper first with warm water, then with alcohol (?), and after careful drying, weigh. What is electro-plating?

SILVER.—Ag.

(1) *Prepare Pure Silver from a Coin.*

Dissolve a 25-cent piece in nitric acid of specific gravity 1.2, filter (?), and evaporate the blue (?) solution to dryness. Fuse the residue till it blackens, extract with 250 cc. of water; filter. Now add ammonia in large excess, and then, cautiously, a sodium bisulphite solution (of about 40 per cent.) until on boiling a small portion of the liquid, it is completely decolorized.

The greater part of the silver separates from the solution on standing in the cold; it is well crystallized. The remainder may be precipitated by warming to 70°. Digest the product with strong ammonia (?), wash, dry and ignite it.

Examine the metal carefully. What are its physical and chemical characteristics?

SILVER AND SULPHUR.

(2) *Silver Sulphide.*—Ag_2S.

Into a dilute solution of silver nitrate (see next experiment), containing about two grams of the metal, pass hydrogen sulphide. When the liquid smells of the gas, filter off the black precipitate, wash it with water and dry at 100°.

Salts.

(3) *Silver Nitrate.*—$AgNO_3$.

Dissolve the silver obtained in (1) in dilute nitric acid and evaporate to dryness on the water bath. Dissolve the residue in 80 cc. of distilled water, and preserve the solution in a dark bottle (?). What is its reaction with litmus?

Reactions.

(4) 1. Compounds of silver on charcoal before the blowpipe give a white metallic globule (?). 2. To a silver solu-

tion—use the nitrate—add hydrochloric acid. Collect the precipitate on a small filter, wash, dissolve it in ammonia, and add an excess of nitric acid to the solution (?). Explain these reactions. 3. Expose a small portion of the chloride to direct sunlight. Any change? What practical application is made of this reaction? (Read Richter, p. 344.)*

(5) Place strips of the metals zinc, iron, tin, lead, and cadmium in a solution of silver nitrate. What is the result in each case? Explain.

GOLD.—Au.

(1) Prepare pure gold from the commercial metal. Fragments of jewelry or a gold coin may be used. A weighed quantity of material is placed in a small flask and covered with concentrated hydrochloric acid. Heat is applied, while small quantities of nitric acid are continually added from time to time. The resulting solution is evaporated on the water-bath, care being taken to exclude dust. Dissolve the residue in water, filter (?), and add a large excess of a ferrous chloride solution to the filtrate. Upon heating, metallic gold separates as a reddish-brown powder. Allow to settle and decant. Extract repeatedly with dilute, boiling hydrochloric acid, and collect the gold on a filter. Incinerate the latter in a porcelain crucible and weigh the product. How could you distinguish the metal gold from mercury, silver, and copper?

Reactions.

(2) 1. Dissolve a small piece of gold (or of a substance containing gold) in aqua regia, concentrate the solution at a gentle heat and pour it into a porcelain dish. Add a solution of ferric chloride to a stannous chloride solution until the latter is permanently yellow. After diluting, dip a glass rod

---

*If practicable, the instructor should here show and explain the preparation of a photographic negative.

into this and then into the gold solution. A purple streak (purple of Cassius) is formed. 2. Add ferrous sulphate to some of the gold chloride solution (?).

In what respects do the members of this group differ from each other, and how can they be distinguished from the metals of the preceding groups?

Problems.—0.5 gram of mercuric oxide gave on ignition with carbon 0.463 gram of metallic mercury; the specific gravity of the vapor of mercuric chloride referred to hydrogen was found to be 135.5. What is the atomic weight of mercury? 2. The molecule of mercury contains how many atoms, if the vapor density equals 100? 3. On analysis a chalcocite was found to contain 20.15 per cent. of sulphur and 79.85 per cent. of copper. Deduce the molecular formula of the mineral. 4. What quantities of silver, gold, and mercury can be precipitated from their respective solutions by one gram of copper?

CHAPTER XII.

ALUMINIUM, TIN, LEAD, BISMUTH.

ALUMINIUM.—Al.

(1) By what methods is this metal obtained on a large scale? What are its properties? Try the action of the following reagents upon aluminium: hydrochloric acid, nitric acid, and sodium hydroxide solution. Write the reactions.

Salts.

(2) *Sulphate of Aluminium and Potassium.*—$KAl(SO_4)_2 + 12H_2O$.

Prepare saturated solutions of aluminium sulphate and potassium sulphate ; mix these so that the resulting liquid contains the two sulphates approximately in the proportion of their molecular weights. The double sulphate crystallizes on standing. Why? Recrystallize it from water. What is the form of the crystals?

What is an *alum ?* (See Richter, p. 355.)

Reactions.

(3) Use alum. 1. Heat a little of the salt on a platinum wire in the oxidizing flame, moisten with cobalt nitrate, and heat again. A blue mass (?) is the product. 2. To an aqueous solution add ammonia (?). 3. Add ammonium sulphide to another portion of the solution. What do you observe? 4. To the diluted solution add sodium hydroxide, drop by drop. Note the successive changes (?).

TIN.—Sn.

(1) Examine a bar of this metal. 1. Note the sound it emits on bending (?). 2. Etch a smooth surface with hydrochloric acid (?). 3. Try the solution of tin in hot hydrochloric acid. 4. What action have moderately dilute, and concentrated, nitric acid upon it? Write the reactions.

(2) *Determine the Specific Heat of Tin.*

A thin glass beaker of about 200 cc. capacity is carefully covered on the outside with a moderately thin layer of cotton wool. This may be called the *calorimeter.* Pour 100 cc. of distilled water into the beaker. Suspend the thermometer in the water. Place 25 grams of granulated tin into a test-tube, close the mouth of the latter with a plug of cotton. Introduce the test-tube with its contents into a beaker glass containing boiling water. A stout copper wire will serve as a handle. After 10 or 15 minutes the tin will have acquired the temperature of the boiling water—100°. The tube is then

rapidly removed from the latter and its outer surface freed from moisture by quickly passing a towel over it. Remove the cotton from the mouth, and transfer the tin to the calorimeter. While the metal is being introduced raise the thermometer from the water, and replace it as soon as all the metal has been added; stir the liquid well and observe, as accurately as possible, the highest point reached by the mercury column. Approximate results can be obtained from these data. Calculate as follows :—

Let y = temperature of water before introducing the tin.
" z =   "   "   " after "
" w = weight of the water.
" v =   " of the metal.
" x = sp. heat—then

$$x = \frac{100\,(z-y)}{25\,(100-z)}$$

(Study Richter pp. 260–263.) Would the specific heat found for tin, when divided into the constant 6.4 give the same value as that found in experiment (3) for the equivalent of tin? Explain. How many series of tin compounds are there?

(3) *Determine the Equivalent Weight of Tin.*

Place about two grams of tin in a porcelain crucible that has been previously weighed. Cover the metal with 5–10 cc. of concentrated nitric acid. Then carefully apply heat by means of an iron plate. The tin is dissolved, while fumes of nitrogen dioxide are set free. When the acid has been entirely expelled, heat the crucible with the white stannic oxide over a Bunsen burner; allow to cool and weigh.

Let w = weight of crucible and tin dioxide.
" v =   "   "   "   " metallic tin.
" y =   "   "   "
Then $w - v$ = weight of oxygen,
and $v - y$ =   "   " tin.

$$\text{Equiv. of tin} = \frac{(v-y) \times 8}{w-v}$$

Salts.

(4) Stannous Chloride.—$SnCl_2$.

Dissolve 10 grams of granulated tin in warm concentrated hydrochloric acid with the addition of a few drops of platinic chloride (?). Put the solution into a well-stoppered bottle.

Reactions.

(5) Stannous Compounds.—Use the chloride solution.

1. Conduct hydrogen sulphide through a portion of the diluted liquid. A brown precipitate (?) is thrown down. Is it soluble in yellow ammonium sulphide? What does hydrochloric acid precipitate from the sulphide solution? 2. What is the action of mercuric chloride upon stannous chloride (see p. 64)?

(6) Stannic Compounds.—Add a few drops of bromine to a portion of the stannous chloride solution, and boil (?). Use the diluted liquid for the tests.

1. Pass hydrogen sulphide into a portion of the solution. What is the color of the precipitate? Is it soluble in hydrochloric acid? in ammonium sulphide?

2. Add copper-turnings, boil, decant the liquid, and add mercuric chloride. What happens? Explain.

LEAD.—Pb.

(1) How can this metal be obtained from the oxide? By what physical properties can it be distinguished from other metals? Is it soluble in the mineral acids? (2) In a solution of five grams of lead acetate in about 50 cc. of water, suspend a strip of metallic zinc and let stand for a few days (?).

Salts.

(3) Dissolve five grams of granulated lead (test-lead) by warming with dilute nitric acid. Concentrate by evaporation and allow to crystallize.

Reactions.

(4) 1. Before the blowpipe, on charcoal, lead compounds are reduced to metallic beads, which are sectile with the knife. 2. Add hydrochloric acid to a solution of the nitrate. Boil the precipitate with water (?). What takes place on cooling? 3. To another portion add dilute sulphuric acid (?). 4. Pass hydrogen sulphide into a third portion (?).

BISMUTH.—Bi.

Reactions.

(1) 1. Mix a little of the oxide or nitrate of bismuth with sodium carbonate and heat in the reducing flame on charcoal. Does the resulting metallic globule resemble lead? Is it sectile? 2. Pass hydrogen sulphide into a solution of the chloride or nitrate in presence of hydrochloric acid (?). 3. Add a large volume of water to a bismuth solution. What occurs? What reactions distinguish aluminium, tin, lead, and bismuth from each other and from the metals previously studied?

Problems.—1. What is the molecular formula of a mineral containing

$$
\begin{aligned}
SiO_2 &= 43.08 \\
Al_2O_3 &= 36.82 \\
CaO &= 20.10 \\
\hline
&100.00
\end{aligned}
$$

2. A compound of tin and chlorine yielded on analysis 29.42 parts of tin and 35.40 parts of chlorine; its vapor density was ascertained to be 132.85. What is the atomic weight of tin? 3. Deduce the formula of *Cosalite* from the following analysis:—

$$
\begin{aligned}
S &= 15.27 \\
Bi &= 41.76 \\
Pb &= 40.32 \\
Ag &= 2.65 \\
\hline
&100.00
\end{aligned}
$$

CHAPTER XIII.

CHROMIUM, MANGANESE, IRON, NICKEL, COBALT.

CHROMIUM.—Cr.

CHROMIUM AND OXYGEN.

(1) *Chromic Oxide.*—Cr_2O_3.

1. *Preparation.*—Mix intimately 20 grams of potassium dichromate and four grams of sulphur. Heat the mixture in a porcelain crucible over the blast lamp for about 20 minutes. Cool, extract the residue with boiling water and dry it at a gentle heat. What is its color; is it soluble in dilute hydrochloric acid? 2. Fuse a portion of it with six times its weight of sodium bisulphate in a platinum crucible. What takes place? 3. Repeat this experiment with some finely powdered chromite. (?)

Salts.

(2) *Chromic Chloride.*—$CrCl_3$.

Prepare the Anhydrous Salt.—Intimately mix 10 grams of chromic oxide, prepared as described, and three grams of powdered charcoal, and convert this into a dough with a little starch paste. Form the product into balls of the size of a pea; dry, and then ignite these (covered with charcoal powder) in a Hessian crucible, provided with well-fitting lid. Place the residue into a tube of hard glass, and heat it in a current of carbon dioxide to expel every trace of moisture. With the aid of a blast lamp increase the temperature and replace the carbon dioxide by a current of chlorine. The excess of chlorine should be absorbed by conducting it into a bottle filled with caustic soda. (?) The resulting chromium trichloride sublimes to the cooler portions of the tube. Describe its appearance. Is it soluble in water?

What other chlorides are prepared in a similar way? Write the equation, expressing the reaction.

(3) *Chrome Alum.*—$Cr_2(SO_4)_3.K_2SO_4 + 24\,H_2O$.

Dissolve 10 grams of potassium bichromate in a little water; acidify with sulphuric acid, pass sulphur dioxide into the liquid until the latter is saturated with the gas. Allow to stand; the double salt crystallizes. What is its crystalline form? Dissolve some of it in cold water and note the color of the solution; now warm it. What takes place (see Richter, p. 382)?

(4) 1. Examine crystals of *potassium dichromate*. How is it obtained? 2. Dissolve 10 grams of this salt in water, and from a burette carefully add a caustic soda solution until the color is changed to yellow (?). What crystallizes from the solution on evaporation? How can you reconvert the product into the dichromate?

Reactions.

(5) 1. Dissolve a minute quantity of a chromium compound in a borax bead. Heat in the oxidizing and in the reducing flame. Results? 2. Heat a little of the compound with potassium nitrate on a platinum foil (?).

(6) *Chromic Compounds.*—Use chrome alum for the tests. 1. Add caustic soda, drop by drop, to a little of the solution. (?) Continue the addition of the reagent till the precipitate is redissolved. What takes place on boiling the solution? 2. What is the action of ammonia on the solution of the chromium salt? 3. Add an excess of potassium hydrate to the chromium solution until the precipitate is redissolved. Now add 5–10 cc. of bromine water and boil. What occurs? Explain.

(7) *Chromates.*—Use a solution of potassium chromate. 1. Add lead acetate solution. Note the color of the precipitate. Is it soluble in acetic acid? 2. Substitute barium chloride for the lead salt in the preceding experiment (?).

3. Acidify the chromate solution with sulphuric acid and add hydrogen peroxide to the liquid. What happens? 4. To some of the chromate solution add a few drops of hydrochloric acid and about one cc. of alcohol. What occurs when the mixture is heated to boiling?

MANGANESE.—Mn.

MANGANESE AND OXYGEN.

(1) In what proportions do these two elements unite with each other? Enumerate the oxides which occur in nature. What is formed when the oxides of manganese are heated in hydrogen? When they are ignited in the air?

Salts.

(2) *Manganous Chloride.*—$MnCl_2 + 4H_2O$.

Evaporate in a porcelain dish the solution obtained in the preparation of chlorine from manganese dioxide and hydrochloric acid. Heat the dry residue over a small flame for some time. Add much water and boil. Filter, and to $\frac{1}{10}$ of the filtrate add a solution of sodium carbonate in excess. Allow the precipitate (?) to settle, draw off the supernatant liquid with a siphon, and wash the remaining precipitate several times with water by decantation. Add the precipitate then to the principal solution, and digest at a gentle heat until a small filtered sample mixed with ammonium sulphide gives a flesh-colored precipitate which is completely dissolved by dilute acetic acid. Now filter and evaporate to crystallization.

(3) *Potassium Manganate*—K_2MnO_4 and *Potassium Permanganate*—$K_2Mn_2O_8$.

In a porcelain crucible fuse a mixture of five grams caustic potash and 2.5 grams potassium chlorate; gradually add five grams finely powdered manganese dioxide. Maintain a moderate red heat for 15 minutes. Dissolve the dark-green

residue in a *little* water. Observe the color of the solution. What does it contain? Then dilute with much water and conduct carbon dioxide into the liquid. Is there any change? If so, write the equation expressing it.

Potassium permanganate as well as potassium manganate are powerful oxidizing agents. **Warm** a little of the alkaline potassium manganate solution with a few drops of alcohol (?). To a little of the permanganate solution, acidified with sulphuric acid, add sulphurous acid (?). Treat the acidified solution also with solutions of ferrous sulphate and oxalic acid (?).

Reactions.

(4) 1. What color do manganese compounds impart to a borax bead in the oxidizing flame? What is the effect of the reducing flame? 2. Heat a little of a manganese compound with sodium carbonate and potassium nitrate on a platinum foil. What does the resulting mass contain? 3. To a little of the solution of the chloride in water add ammonium sulphide. What is the color of the precipitate? Test its solubility in acids (including acetic acid). 4. Add caustic soda to another portion of the chloride solution. Is the precipitate soluble in an excess of the reagent? Is its color affected by exposure to the air? Explain.

IRON.—Fe.

(1) *Preparation.*—Into a tube of Bohemian glass place a porcelain boat filled with the finely powdered oxide. Pass a current of dry hydrogen through the tube, and when all the air is expelled (how could you test it?), apply heat to that part of the tube which contains the boat. What is formed in the anterior portion of the tube! After a red heat has been maintained for 10 minutes allow the boat to cool in hydrogen and examine its contents. Are they attracted by the magnet? Expose the product to the air (?).

How is iron obtained from its ores on a large scale? What are its properties? (see Richter, pp. 396, 400). Distinguish between cast-iron, steel, and wrought-iron.

Salts.

(2) *Ferrous Sulphate.*—$FeSO_4 + 7H_2O$.

To 25 grams of iron in the form of nails or wire, free from rust, contained in a flask, add 200 cc. of dilute (1 : 4) sulphuric acid. When the evolution of the gas (? Note its odor!) is no longer violent, warm, and finally boil until the liberation of gas ceases. A sample of the solution poured into a test tube should, on cooling, give a copious separation of crystals. Filter into a casserole containing two to three cc. of conc. sulphuric acid, and let stand for eight hours. Collect the crystallized product in a funnel the stem of which is closed with a loose plug of glass wool,* allow the mother liquor to drain off, wash with very little cold water (?), and dry between sheets of filter paper. Examine the product carefully. Note its color, taste, solubility in water, and crystal form. What other salts of analogous composition are isomorphous with it?

What is observed when some of the salt is heated, first moderately, then strongly, in a tube of hard glass?

Expose the aqueous solution of the salt to the air for several hours (?).

(3) *Ferrous Ammonium Sulphate.*—$Fe(NH_4)_2(SO_4)_2 + 6H_2O$.

In 100 cc. of dilute sulphuric acid dissolve clean iron wire till no more hydrogen is given off; neutralize a like quantity of the acid exactly with ammonia water, and add to it a few drops of dilute sulphuric acid. Filter the iron solution into that of

---

* It is better to use a perforated platinum cone, and to remove the adhering solution with the aid of a filter pump.

the ammonium salt. Let the salt crystallize, drain it on a funnel provided with a perforated platinum cone, wash and dry as described under (2). Preserve in a well-stoppered bottle. What metals can replace the iron in this salt without altering its crystalline form?

(4) *Ferric Ammonium Sulphate.*—$Fe_2(SO_4)_3 \cdot (NH_4)_2SO_4 + 24H_2O$.

Place 20 grams of crystallized ferrous sulphate into a porcelain dish together with a few cc. of water and 3.5 grams of oil of vitriol. Warm on an asbestos plate, adding nitric acid, drop by drop, until no further change of color (?) is observed. Evaporate the excess of nitric acid, dissolve the residue in hot water and add 3.5 grams of ammonium sulphate; filter, and set the solution aside for crystallization. Separate the crystals from the mother liquor, and wash and dry them as under (2). To what class of substances does this salt belong? Why?

Reactions.

(5) In a borax bead dissolve a small quantity of an iron compound, and treat it successively in the oxidizing and reducing flames. What changes do you observe?

(6) *Ferrous Compounds.*—Use a *freshly* prepared solution of ferrous sulphate for the following tests: 1. To a few drops of it, diluted with water, add ammonia. Note the color of the precipitate, and the changes which occur on exposure to the air (?). 2. Add ammonium sulphide to another portion (?). Is the resulting precipitate soluble in hydrochloric acid? 3. In a porcelain capsule bring together a little of the ferrous solution and a drop of a potassium ferrocyanide solution. Result? 4. In a similar manner test a drop of the iron solution with ferricyanide of potassium.

(7) *Ferric Compounds.*—In the presence of free acids, oxidizing agents convert iron compounds from the ferrous into the ferric condition. 1. Acidify the ferrous sulphate solution with sulphuric acid, warm, and add concentrated

nitric acid until it fails to produce a change in color: the iron is then in the ferric state. 2. Dilute a few drops of the yellow liquid with several cc. of water and add ammonia (?). 3. Test a drop of the ferric solution with potassium ferrocyanide (?). 4. Treat a second drop with ferricyanide of potassium (?). 5. Mix another drop with a solution of potassium sulphocyanate (?). 6. Conduct hydrogen sulphide into some of the ferric sulphate solution. What do you observe? Explain the reaction, and write the equation expressing it. 7. Place a piece of metallic zinc in a test-tube containing a solution of the ferric salt. What takes place?

(8) *Quantitative Estimation of Iron.*—Under manganese it was observed that the salt potassium permanganate is an oxidizing agent. To show how this salt acts with iron in its lower form of oxidation, fill a burette with an aqueous solution of it; allow it to drop slowly into the solution of a ferrous salt acidulated with sulphuric acid. The pink color of the permanganate immediately disappears on stirring with a glass rod. This continues until the ferrous salt is completely oxidized to the ferric state. A drop of permanganate added in excess will then impart a faint pink color to the liquid. This *indicates* that the reaction is ended. Write the equation.

FIG. 40.

This behavior may be utilized for determining the *quantity* of iron in a solution. That this may be done, it is first necessary to standardize the permanganate solution. Proceed as follows: Dissolve about two grams of the permanganate in 1000 cc. of water. Fill a burette with this solution. Weigh out .2 gram of clean piano wire. Place this into a small flask (Fig. 40) provided with a cork and valve.* Cover the iron wire with dilute sulphuric acid. Warm. When the iron is

---

*With a sharp knife make a longitudinal incision of about one cm, length, in a rubber tube, and close one end by means of a glass rod,

·completely dissolved, remove the cork, add cold water to the solution, and slowly admit the permanganate until the final pink coloration appears. Note the volume of permanganate required to produce this effect. Suppose 30 cc. had been consumed, then :—

$$30 \text{ cc. } K_2Mn_2O_8 = .2000 \text{ gram metallic iron.}$$
$$1 \text{ `` } \text{ `` } = .00666 \text{ `` } \text{ `` } \text{ ``}$$

This is then the standard of the permanganate in iron units.

Next, dissolve one gram of ferrous ammonium sulphate, in 100 cc. distilled water, add five cc. of sulphuric acid, and then introduce the permanganate until the final reaction is observed. Calculate the percentage of iron in this salt and compare the experimental result with the theoretical value.

How much oxygen will each molecule of permanganate give up in oxidizing? How many molecules of ferrous oxide can be changed to ferric oxide by a molecule of potassium permanganate.

COBALT.—Co, AND NICKEL.—Ni.

Reactions.

1. Dissolve a minute quantity of a cobalt compound in a borax bead. Heat first in the oxidizing, then in the reducing flame (?). 2. What is the behavior of nickel compounds under like conditions? 3. Add caustic alkali to a solution of cobalt nitrate, warm the mixture (?). 4. What action have caustic alkalies on solutions of nickel salts? 5. To the cobalt solution cautiously add ammonia. After a precipitate (?) has formed, add more of the reagent. What takes place? Expose the resulting solution to the air in a shallow dish (?). 6. Treat a nickel solution in an analogous manner (?) 7. To the solutions of cobalt and nickel each in a separate test-tube, add ammonium sulphide. Filter and wash the precipitated sulphides, and test their solubility in acids (?).

Note the colors of cobalt and nickel salts, in the hydrated as well as in the anhydrous state.

Nickel Carbonyl.—A tube of hard glass is drawn out to a bayonet at one end, and filled with nickel oxide. A current of hydrogen is conducted through the tube, while the temperature is raised to a low red heat. This is maintained until moisture ceases to condense, when the escaping hydrogen impinges on a cool surface. Cool in an atmosphere of hydrogen. Carbon monoxide, washed with caustic potash and sulphuric acid, is now passed through the apparatus, and the issuing gas is led through a tube (Fig. 41) surrounded by a freezing mixture (care!). What is the colorless mobile liquid which condenses? This experiment should be conducted under a hood (?).

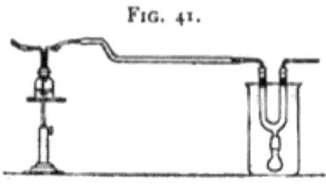

Fig. 41.

Is there any marked difference between cobalt and nickel in respect to their chemical deportment?

Point out the differences in the reactions of chromium, manganese, iron, cobalt, and nickel.

How may ferrous compounds be distinguished from ferric? What conditions are favorable to the conversion of the former into the latter? The latter into the former?

By what means may chromic salts be changed into compounds of chromic acid? How may the reverse change be effected?

Devise a method for separating the elements treated in this chapter.

Problems.—1. How much potassium bichromate can be obtained theoretically from 100 kilos of a chromite containing 58.6 per cent. of chromic oxide? 2. 100 grams of a

pyrolusite, which was found to contain four per cent. of impurities, will give what volume of oxygen, measured at 20° C. and 745 mm., when strongly ignited? What is the weight of the residue, assuming that one-half of the impurities was moisture, the other half quartz? 3. How many grams of nitric acid are required to oxidize 12 grams of crystallized ferrous sulphate? 4. What percentage of metallic iron is contained in a salt, of which .7 gram are exactly oxidized by 17.8 cc. of permanganate solution (standard: 1 cc. = .0056 gram Fe)?

CHAPTER XIV.

Name the metals constituting the platinum group.

PLATINUM.

(1) Examine the metal in the form of foil or wire. Note its color and lustre. Determine its specific gravity. Is it fusible in the flame of the Bunsen burner or the blast lamp?

(2) What are its chemical properties? Place a small piece of the metal into a test-tube and heat it with concentrated hydrochloric acid. Any action? Treat another piece with nitric acid. Result? Combine the contents of both test-tubes and warm. What takes place?

Platinic Chloride—$PtCl_4$.—Boil platinum scrap with hydrochloric acid. Decant the liquid, and dissolve the remaining metal in nitro-hydrochloric acid. Concentrate the resulting solution by evaporation on the water bath, and add a strong solution of ammonium chloride. Filter off the precipitated platinic ammonium chloride, and after careful drying, ignite it in a large porcelain crucible. Extract the residue or spongy platinum with boiling hydrochloric acid, then dissolve it in *aqua regia*, and evaporate to dryness on the water bath. Moisten the residue with hydrochloric acid and again evaporate. Dissolve the product in distilled water.

Reactions.

Use platinic chloride for the tests. 1. Add hydrogen sulphide water and apply a gentle heat. 2. Add a little grape sugar and caustic soda to a second portion. Warm. See under potassium and sodium how solutions of this metal behave with alkaline chlorides, etc.

APPENDIX.

TABLE OF METRIC WEIGHTS AND MEASURES.

MEASURES OF LENGTH.

1 metre = 10 decimetres = 100 centimetres = 1000 millimetres.
1 metre = 1.09363 yards = 3.2809 feet = 39.3709 inches.

MEASURES OF CAPACITY.

1 cubic metre = 1000 litres = 1,000,000 cubic centimetres 1,000,000,000 cubic millimetres.

1 litre = 61.02705 cubic inches = .035317 cubic foot = 1.76077 pints = .22097 gallon.

MEASURES OF WEIGHT.

1 gram = weight of 1 cc. of water at 4° C.

1 Kilogram = 1000 grams = 100.000 centigrams = 1,000,000 milligrams.

1 Kilogram = 2.20462 lbs. = 35.2739 ounces = 15432.35 grains.

TABLE OF ATOMIC WEIGHTS OF ELEMENTS.

Aluminium,	Al	27.0	Chromium,	Cr	52.5
Antimony,	Sb	120.0	Cobalt,	Co	59.0
Arsenic,	As	75.0	Copper,	Cu	63.3
Barium,	Ba	137.0	Fluorine,	Fl	19.0
Bismuth,	Bi	208.0	Gold,	Au	197.0
Boron,	B	11.0	Hydrogen,	H	1.0
Bromine,	Br	80.0	Iodine,	I	127.0
Cadmium,	Cd	112.0	Iron,	Fe	56.0
Calcium,	Ca	40.0	Lead,	Pb	207.0
Carbon,	C	12.0	Magnesium,	Mg	24.0
Chlorine,	Cl	35.5	Manganese,	Mn	55.0

TABLE OF ATOMIC WEIGHTS OF ELEMENTS.—*Continued.*

Mercury,	Hg	200.0	Silicon,	Si	28.0
Molybdenum,	Mo	96.0	Silver,	Ag	108.0
Nickel,	Ni	59.0	Sodium,	Na	23.0
Nitrogen,	N	14.0	Strontium,	Sr	87.5
Oxygen,	O	16.0	Sulphur,	S	32.0
Phosphorus,	P	31.0	Tin,	Sn	118.0
Platinum,	Pt	195.0	Zinc,	Zn	65.0
Potassium,	K	39.0			

TENSION OF AQUEOUS VAPOR IN MILLIMETRES (*Regnault*).

Temp.	Tension.	Temp.	Tension.	Temp.	Tension.
$0°$	4.6	$11°$	9.8	$21°$	18.5
1	4.9	12	10.4	22	19.7
2	5.3	13	11.1	23	20.9
3	5.7	14	11.9	24	22.2
4	6.1	15	12.7	25	23.6
5	6.5	16	13.5	26	25.0
6	7.0	17	14.4	27	26.5
7	7.5	18	15.4	28	28.1
8	8.0	19	16.3	29	29.8
9	8.5	20	17.4	30	31.6
10	9.1				

THE SPECIFIC GRAVITY AND THE WEIGHT OF A LITRE OF GASES.

	SPECIFIC GRAVITY (*Regnault*).		The Weight of one Litre in its Normal Condition, in Grams.
	At $0°$ and 760 mm. Referred to Water at $4°$.	Referred to Air Under like Temperature and Pressure.	
Air,	0.0012928	1.00000	1.2932
Oxygen,	0.0014293	1.10563	1.4300
Nitrogen,	0.0012557	0.97137	1.2562
Hydrogen,	0.00008954	0.06926	0.0896
Carbon dioxide,	0.001977	1.529	1.9663

www.ingramcontent.com/pod-product-compliance
Lightning Source LLC
Chambersburg PA
CBHW030308170426
43202CB00009B/919